ENVIRONMENTAL REMEDIATION TECHNOLOGIES, REGULATIONS AND SAFETY

CLIMATE AND ENVIRONMENTAL PROTECTION

INTERNATIONAL FUNDING

ENVIRONMENTAL REMEDIATION TECHNOLOGIES, REGULATIONS AND SAFETY

Additional books in this series can be found on Nova's website under the Series tab.

Additional e-books in this series can be found on Nova's website under the e-book tab.

ENVIRONMENTAL REMEDIATION TECHNOLOGIES, REGULATIONS AND SAFETY

CLIMATE AND ENVIRONMENTAL PROTECTION

INTERNATIONAL FUNDING

JACKSON M. GARCIA
EDITORS

New York

For permission to use material from this book please contact us:
Telephone 631-231-7269; Fax 631-231-8175
Web Site: http://www.novapublishers.com

Library of Congress Cataloging-in-Publication Data

ISBN: 978-1-63117-960-0

Published by Nova Science Publishers, Inc. † New York

CONTENTS

Preface		**vii**
Chapter 1	International Climate Change Financing: The Climate Investment Funds (CIFs) *Richard K. Lattanzio*	**1**
Chapter 2	International Climate Change Financing: The Green Climate Fund (GCF) *Richard K. Lattanzio*	**27**
Chapter 3	International Environmental Financing: The Global Environment Facility (GEF) *Richard K. Lattanzio*	**47**
Index		**77**

PREFACE

Many governments acknowledge that environmental degradation and climate change pose international and trans-boundary risks to human populations, economies, and ecosystems that could result in a worsening of poverty, social tensions, and political stability. To confront these global challenges, countries have negotiated various international agreements to protect the environment, reduce pollution, conserve natural resources, and promote sustainable growth. While some observers have called upon developed countries to take the lead in addressing these issues, efforts are unlikely to be sufficient without similar measures being implemented in developing countries. Developing countries, however, focused on poverty reduction and economic growth, may not have the financial resources, technological know-how, or institutional capacity to deploy such measures. Therefore, international support for these areas has remained the principal method for governments to assist developing country action on global environmental problems. This book focuses on international funding for climate and environmental protection through the Climate Investment Funds, the Green Climate Fund, and the Global Environment Facility.

Chapter 1 – The United States contributes funding to various international financial institutions to assist developing countries to address global climate change and other environmental concerns. Congress is responsible for several activities in this regard, including (1) authorizing periodic appropriations for U.S. financial contributions to the institutions, and (2) overseeing U.S. involvement in the programs. Issues of congressional interest include the overall development assistance strategy of the United States, U.S. leadership in global environmental and economic affairs, and U.S. commercial interests in trade and investment. This report provides an overview of two of the larger and more recently instituted international financial institutions for the environment—the Climate Investment Funds (CIFs)—and analyzes their structure, funding, and objectives in light of the many challenges within global environmental finance.

The CIFs are investment programs administered by the multilateral development banks (MDBs) that aim to help finance developing countries' transitions toward low-carbon and climate-resilient development. Formally approved by the World Bank's Board of Directors on July 1, 2008, the CIFs are composed of two trust funds—the Clean Technology Fund (CTF) and the Strategic Climate Fund (SCF)—each with a specific scope, objective, and governance structure. The CTF provides financing for demonstrating, deploying, and diffusing low-carbon technologies that have the potential for long-term avoidance of greenhouse gas emissions. The SCF—a suite of three separate funds, including the Pilot Program for Climate Resilience (PPCR), the Forest Investment Program (FIP), and the Scaling Up Renewable Energy Program in Low Income Countries (SREP)—supports the least developed countries in their efforts to achieve low-carbon, climate-resilient development. Overall, donor countries have pledged $7.6 billion to the funds since September 2008 in support of programs in 49 developing countries. The U.S. pledge in 2008 was for a total of $2 billion. For FY2010, Congress approved $375 million for the CIFs (the Consolidated Appropriations Act, 2010, H.R. 3288; P.L. 111-117); for FY2011, Congress approved $234.5 million (the Department of Defense and Full-Year Continuing Appropriations Act, 2011, H.R. 1473; P.L. 112-10); for FY2012, Congress approved $234.5 million (the Consolidated Appropriations Act, 2012, H.R. 2055; P.L. 112-74); and for FY2013, Congress approved $234.5 million (the Consolidated and Further Continuing Appropriations Act, 2013, H.R. 933; P.L. 113-6). For FY2014, the Administration requested $283.7 million for the funds.

The CIFs are just one set of financial mechanisms in a larger network of international programs designed to address the global environment. Accordingly, their effectiveness depends on how the funds address programmatic issues, build upon national investment plans, react to recent developments in the financial landscape, and respond to emerging opportunities. Proponents of the CIFs point to several factors in support of the funds, including an innovative programmatic design, a country-led investment process, and a balanced governance structure with enhanced stakeholder engagement. Proponents of the MDBs' role in environmental assistance emphasize several advantages to financing climate programs through the MDBs, including its commitment to private sector development, its capacity to leverage large co-financing arrangements, and its possession of fiduciary standards and institutional expertise. However, critics highlight several factors of concern with the CIFs and their Trustee, including a lack of transparency, coordination, and "polluter pay" responsibilities; a potential for increased debt burdens on developing countries; and a prior economic development policy at the development banks that is considered a conflict of interest for environmental protection.

Chapter 2 – Over the past several decades, the United States has delivered financial and technical assistance for climate change activities in the developing world through a variety of bilateral and multilateral programs. The United States and other industrialized countries committed to such assistance through the United Nations Framework Convention on Climate Change (UNFCCC, Treaty Number: 102-38, 1992), the Copenhagen Accord (2009), and the UNFCCC Cancun Agreements (2010), wherein the higher-income countries pledged jointly up to $30 billion of "fast start" climate financing for lower-income countries for the period 2010-2012, and a goal of mobilizing jointly $100 billion annually by 2020. The Cancun Agreements also proposed that the pledged funds are to be new, additional to previous flows, adequate, predictable, and sustained, and are to come from a wide variety of sources, both public and private, bilateral and multilateral, including alternative sources of finance.

One potential mechanism for mobilizing a share of the proposed international climate financing is the UNFCCC Green Climate Fund (GCF), proposed in the Cancun Agreements and accepted by Parties during the December 2011 conference in Durban, South Africa. The fund aims to assist developing countries in their efforts to combat climate change through the provision of grants and other concessional financing for mitigation and adaptation projects, programs, policies, and activities. The GCF is to be capitalized by contributions from donor countries and other sources, including both innovative mechanisms and the private sector. Currently, the GCF complements many of the existing multilateral climate change funds (e.g., the Global Environment Facility, the Climate Investment Funds, and the Adaptation Fund); however, as the official financial mechanism of the UNFCCC, some Parties believe that it may eventually replace or subsume the other funds. While many Parties expect capitalization and operation of the GCF to begin shortly after the November 2013 conference in Warsaw, Poland, many issues remain to be clarified, and some involve long-standing and contentious debate. They include what role the CGF would play in providing sustained finance at scale, how it would fit into the existing development assistance and climate financing architecture, how it would be capitalized, and how it would allocate and deliver assistance efficiently and effectively to developing countries.

The U.S. Congress—through its role in authorizations, appropriations, and oversight—would have significant input on U.S. participation in the GCF. Congress regularly determines and gives guidance to the allocation of foreign aid between bilateral and multilateral assistance as well as among the variety of multilateral mechanisms. In the past, Congress has raised concerns regarding the cost, purpose, direction, efficiency, and effectiveness of the UNFCCC and existing international institutions of climate financing. Potential authorizations and appropriations for the GCF would rest with several committees, including the U.S. House of Representatives Committees on Foreign Affairs (various subcommittees); Financial Services (Subcommittee on International Monetary Policy and Trade); and Appropriations (Subcommittee on State, Foreign Operations, and Related Programs); and the U.S. Senate Committees on Foreign Relations (Subcommittee on International Development and Foreign Assistance, Economic Affairs, and International Environmental Protection); and Appropriations (Subcommittee on State, Foreign Operations, and Related Programs). As of April 2013, the U.S. Administration—through its State, Foreign Operations, and Related Programs 150 account—has made no specific budget request for appropriated funds to be contributed to the GCF.

Chapter 3 – This report provides an overview of one of the oldest international financial institutions for the environment—the Global Environment Facility (GEF)—and analyzes its structure, funding, and objectives in light of the many challenges within the contemporary landscape of global environmental finance.

GEF is an independent and international financial organization that provides grants, promotes cooperation, and fosters actions in developing countries to protect the global environment. Established in 1991, it unites 182 member governments and partners with international institutions, nongovernmental organizations, and the private sector to assist developing countries with environmental projects related to six areas: biodiversity, climate change, international waters, the ozone layer, land degradation, and persistent organic pollutants. GEF receives funding from multiple donor countries—including the United States—and provides grants to cover the additional or "incremental" costs associated with transforming a project with national benefits into one with global environmental benefits. In this way, GEF funding is structured to "supplement" base project funding and provide for the environmental components in national development agendas. GEF partners with several international agencies, including the International Bank for Reconstruction and Development, the United Nations Development Program (UNDP), and the United Nations Environment Program (UNEP), among others, and is the primary fund administrator for four Rio (Earth Summit) Conventions, including the Convention on Biological Diversity (CBD), the United Nations Framework Convention on Climate Change (UNFCCC), the Stockholm Convention on Persistent Organic Pollutants (POPs), and the United Nations Convention to Combat Desertification (UNCCD). GEF also establishes operational guidance for international waters and ozone activities, the latter consistent with the Montreal Protocol on Substances that Deplete the Ozone Layer and its amendments. Since its inception, GEF has allocated $11.5 billion—supplemented by more than $57 billion in cofinancing—for more than 3,200 projects in over 165 countries.

GEF is one mechanism in a larger network of international programs designed to address the global environment. Accordingly, its effectiveness depends on how the fund addresses programmatic issues, builds upon national investment plans, reacts to recent developments in the financial landscape, and responds to emerging opportunities. Critics contend that the existing system has had limited impact in addressing major environmental concerns—specifically climate change and tropical deforestation—and has been unsuccessful in delivering global transformational change. A desire to achieve more immediate impacts has led to a restructuring of the Multilateral Development Banks' (MDBs') role in environmental finance and the introduction of many new bilateral and multilateral funding initiatives. The future of GEF remains in the hands of the donor countries, including the United States, which can choose to broaden the mandate and/or strengthen its institutional arrangements or reduce and replace it by other bilateral or multilateral funding mechanisms.

In: Climate and Environmental Protection
Editor: Jackson M. Garcia
ISBN: 978-1-63117-960-0

Chapter 1

INTERNATIONAL CLIMATE CHANGE FINANCING: THE CLIMATE INVESTMENT FUNDS (CIFS)*

Richard K. Lattanzio

SUMMARY

The United States contributes funding to various international financial institutions to assist developing countries to address global climate change and other environmental concerns. Congress is responsible for several activities in this regard, including (1) authorizing periodic appropriations for U.S. financial contributions to the institutions, and (2) overseeing U.S. involvement in the programs. Issues of congressional interest include the overall development assistance strategy of the United States, U.S. leadership in global environmental and economic affairs, and U.S. commercial interests in trade and investment. This report provides an overview of two of the larger and more recently instituted international financial institutions for the environment—the Climate Investment Funds (CIFs)—and analyzes their structure, funding, and objectives in light of the many challenges within global environmental finance.

---

* This is an edited, reformatted and augmented version of a Congressional Research Service publication, CRS Report for Congress R41302, prepared for Members and Committees of Congress, from www.crs.gov, dated June 3, 2013.

The CIFs are investment programs administered by the multilateral development banks (MDBs) that aim to help finance developing countries' transitions toward low-carbon and climate-resilient development. Formally approved by the World Bank's Board of Directors on July 1, 2008, the CIFs are composed of two trust funds—the Clean Technology Fund (CTF) and the Strategic Climate Fund (SCF)—each with a specific scope, objective, and governance structure. The CTF provides financing for demonstrating, deploying, and diffusing low-carbon technologies that have the potential for long-term avoidance of greenhouse gas emissions. The SCF—a suite of three separate funds, including the Pilot Program for Climate Resilience (PPCR), the Forest Investment Program (FIP), and the Scaling Up Renewable Energy Program in Low Income Countries (SREP)—supports the least developed countries in their efforts to achieve low-carbon, climate-resilient development. Overall, donor countries have pledged $7.6 billion to the funds since September 2008 in support of programs in 49 developing countries. The U.S. pledge in 2008 was for a total of $2 billion. For FY2010, Congress approved $375 million for the CIFs (the Consolidated Appropriations Act, 2010, H.R. 3288; P.L. 111-117); for FY2011, Congress approved $234.5 million (the Department of Defense and Full-Year Continuing Appropriations Act, 2011, H.R. 1473; P.L. 112-10); for FY2012, Congress approved $234.5 million (the Consolidated Appropriations Act, 2012, H.R. 2055; P.L. 112-74); and for FY2013, Congress approved $234.5 million (the Consolidated and Further Continuing Appropriations Act, 2013, H.R. 933; P.L. 113-6). For FY2014, the Administration requested $283.7 million for the funds.

The CIFs are just one set of financial mechanisms in a larger network of international programs designed to address the global environment. Accordingly, their effectiveness depends on how the funds address programmatic issues, build upon national investment plans, react to recent developments in the financial landscape, and respond to emerging opportunities. Proponents of the CIFs point to several factors in support of the funds, including an innovative programmatic design, a country-led investment process, and a balanced governance structure with enhanced stakeholder engagement. Proponents of the MDBs' role in environmental assistance emphasize several advantages to financing climate programs through the MDBs, including its commitment to private sector development, its capacity to leverage large co-financing arrangements, and its possession of fiduciary standards and institutional expertise. However, critics highlight several factors of concern with the CIFs and their Trustee, including a lack of transparency, coordination, and "polluter pay" responsibilities; a potential for increased debt burdens on developing countries; and a prior economic development policy at the development banks that is considered a conflict of interest for environmental protection.

INTRODUCTION

Many governments acknowledge that environmental degradation and climate change pose international and trans-boundary risks to human populations, economies, and ecosystems that could result in a worsening of poverty, social tensions, and political stability. To confront these global challenges, countries have negotiated various international agreements to protect the environment, reduce pollution, conserve natural resources, and promote sustainable growth. While some observers have called upon developed countries to take the lead in addressing these issues, efforts are unlikely to be sufficient without similar measures being implemented in developing countries. Developing countries, however, focused on poverty reduction and economic growth, may not have the financial resources, technological know-how, or institutional capacity to deploy such measures. Therefore, international support for these areas has remained the principal method for governments to assist developing country action on global environmental problems.[1]

The United States and other industrialized countries have committed to financial assistance for environmental initiatives through several multilateral agreements (e.g., the Montreal Protocol (1987), the United Nations Framework Convention on Climate Change (1992), United Nations Convention to Combat Desertification (1994), and the Copenhagen Accord (2009)). International financial assistance takes many forms, from fiscal transfers to market transactions, and includes foreign direct investment (FDI), bilateral overseas development assistance (ODA), and contributions to multilateral development banks (MDB)[2] and other international financial institutions (IFI), as well as the offering of export credits, loan guarantees, and insurance products.

Table 1 outlines recent U.S. financial support for multilateral environmental initiatives. Congress is responsible for several activities in this regard, including (1) authorizing periodic appropriations for U.S. financial contributions to the institutions, and (2) overseeing U.S. involvement in the programs. Issues of congressional interest include the overall development assistance strategy of the United States, U.S. leadership in global environmental and economic affairs, and U.S. commercial interests in trade and investment.[3] As Congress considers potential authorizations and/or appropriations for initiatives administered through the Department of State, the Department of the Treasury, and other agencies with international programs, it may have questions concerning the direction, efficiency, and effectiveness of current bilateral and multilateral programs. This report provides an overview of two of the larger and more recently instituted multilateral mechanisms—the Climate Investment Funds (CIFs)—and analyzes their structure, funding, and objectives in light of the many challenges within the contemporary landscape of global environmental finance.

Table 1. Recent U.S. Budget Authority for Multilateral Climate and Environment Funds In nominal US$ million

Agency/Program	2010 Enacted	2011 Enacted	2012 Enacted	2013 Enacted[a]	2014 Request
Department of State					
Least Developed Country Fund	30.0	25.0	25.0	TBD	TBD
Special Climate Change Fund	20.0	10.0	10.0	TBD	TBD
World Bank Forest Carbon Partnership	10.0	8.0	TBD	TBD	TBD
Department of Treasury					
Tropical Forests Conservation Act	26.0	16.4	12.0	12.0	0.0
Global Environment Facility	86.5	89.8	119.8[b]	129.4[c]	143.8
Climate Investment Fund: Clean Technology Fund	300.0	184.6	229.6[d]	175.3	215.7
Climate Investment Fund:					
Strategic Climate Fund - Pilot	55.0	10.0	18.7[e]	25.0[f]	34.0[g]
Program for Climate Resilience					
Climate Investment Fund:					
Strategic Climate Fund - Forest	20.0	30.0	37.5[e]	12.5[f]	17.0[g]
Investment Program					
Climate Investment Fund:					
Strategic Climate Fund -	0.0	10.0	18.7[e]	12.5[f]	17.0[g]
Scaling-Up Renewable Energy					

Source: Office of Management and Budget, The Budget of the United States Government, 2011, 2012, 2013, and 2014; CRS correspondence with Department of State and Department of the Treasury.

Notes: TBD, "to be determined": Appropriated funds for some programs/activities are drawn from larger line item categories in agency budget authorities, occasionally with "shall"-language implementing spending ceilings. Allocations for these programs are left at the discretion of the agency and have yet to be determined and/or fully reported.

[a] Except where noted, FY2013 Enacted amount is as continuing resolution in the Consolidated and Further Continuing Appropriations Act, 2013 (P.L. 113-6). Figures do not include sequestration reduction.

[b] FY2012 Enacted amount for GEF includes the transfer of $30 million from the Economic Support Fund as provided in the Consolidated Appropriations Act, 2012 (P.L. 112-74).

[c] FY2013 Enacted amount for the GEF is as provided in the Consolidated and Further Continuing Appropriations Act, 2013 (P.L. 113-6).

[d] FY2012 Enacted amount for CTF includes the transfer of $45 million from the Economic Support Fund as provided in the Consolidated Appropriations Act, 2012 (P.L. 112-74).

[e] FY2012 Enacted amount for SCF includes the transfer of $25 million from the Economic Support Fund as provided in the Consolidated Appropriations Act, 2012 (P.L. 112-74).

[f] FY2013 Enacted amount for SCF is $47.3 million for all three programs. The figures in the table reflect Treasury's internal proposal for contribution among the PPCR, FIP, and SREP.

[g] FY2014 Request amount for SCF is $68.0 million for all three programs. The figures above are estimates of contributions to each program. Treasury will finalize contributions among the PPCR, FIP, and SREP in spring 2014.

THE CLIMATE INVESTMENT FUNDS

Background

Projected climate change is considered a potential threat to economic development, with anticipated effects on the environment, human health, food security, and economic activity. Further, climate change disproportionately affects the urban and rural poor of developing countries, thus making it a central concern to those interested in poverty reduction and sustainable development.[4] Under this context, and at the request of the G8/G20, the multilateral development banks (MDBs) have recently sought to expand their support to low-carbon and climate-resilient investments in several ways, including (1) creating new and additional environmental funding resources, (2) repackaging their "core" financial products with specialized climate provisions, and (3) leveraging their suite of financial instruments for greater private sector environmental investment.[5]

In keeping with these aims, in February 2008, Japan, the United Kingdom, and the United States announced their intention to create a set of funds at the MDBs to help developing countries "bridge the gap between dirty and clean energy" and "boost the World Bank's ability to help developing countries tackle climate change."[6] The World Bank held the first design meeting for the proposed Climate Investment Funds (CIFs) in March 2008 in Paris, France. Two subsequent meetings were held in Washington, DC, and Potsdam, Germany, and on May 23, 2008, representatives from 40 developing and industrialized countries reached agreement on the funds' design and duration (the CIFs were programmed to sunset upon the commencement of a new climate fund in the United Nations Framework Convention on Climate Change (UNFCCC)). Formally approved by the World Bank's Board of Directors on July 1, 2008, the CIFs have become an attempt to bridge the gap in climate financing between present obligations and a future global climate change agreement.[7]

The CIFs are composed of two separate trust funds—the Clean Technology Fund (CTF) and the Strategic Climate Fund (SCF)—each with a specific scope, objective, and governance structure. Overall, 14 donor countries have pledged $7.6 billion (in historical value) to the funds since September 2008, which supports programming in 49 developing countries.[8] The U.S. pledge in 2008 was for a total of $2 billion. All U.S. funding is subject to annual congressional approval. Authorizing legislation is managed by the House Financial Services Committee and Senate Foreign Relations Committee. The House and Senate Appropriations Subcommittees on State, Foreign Operations, and Related Programs have jurisdiction over appropriations.

U.S. contributions include the following:

- FY2010, Congress approved $300 million for the CTF and $75 million for the SCF (the Consolidated Appropriations Act, 2010, H.R. 3288; P.L. 111-117).
- FY2011, Congress approved $184.6 million for the CTF and $49.9 million for the SCF (the Department of Defense and Full-Year Continuing Appropriations Act, 2011, H.R. 1473; P.L. 112-10).
- FY2012, Congress approved $184.6 million for the CTF and $49.9 million for the SCF; however, provisions for funding transfers were included. Using these provisions, the Department of State transferred $45 million from its Economic Support Fund to the CTF, and $25 million to the SCF during FY2012 (the Consolidated Appropriations Act, 2012, H.R. 2055; P.L. 112-74).
- FY2013, Congress approved $184.6 million for the CTF and $49.9 million for the SCF through a continuing resolution (the Consolidated and Further Continuing Appropriations Act, 2013, H.R. 933; P.L. 113-6). FY2013-enacted account level estimates are subject to the budget sequestration process as established by the Budget Control Act of 2011 (P.L. 112-25) and the American Taxpayer Relief Act (P.L. 112-240). The total budget impact of sequestration has yet to be determined.

- For FY2014, the Administration has requested \$215.7 million for the CTF and \$68 million for the SCF.[9]

The Clean Technology Fund (CTF)

Overview

Faced with energy and environmental challenges, among others, many developing countries see value in clean technology to meet their energy security, poverty alleviation, and sustainable development goals while also reducing their growth in emissions. However, the costs to developing countries of switching to cleaner technologies without financial assistance may be prohibitive. The CTF seeks to provide financing—principally to larger emerging economies and to regional groups—for demonstrating, deploying, and diffusing low-carbon technologies with the potential for long-term avoidance of greenhouse gas emissions. The fund promotes renewable energy and energy efficient technologies in the power sector as well as energy efficiency strategies in the transportation, building, industry, and agricultural sectors. Currently, the CTF is designed to support 15-20 country and regional investment plans and/or co-financed projects. As of March 2013, the CTF has endorsed 16 investment plans for \$5.58 billion in direct funding (with a projected \$40 billion in leveraged co-financing), including plans from Chile, Colombia, Egypt, India, Indonesia, Kazakhstan, Mexico, Morocco, Nigeria, Philippines, South Africa, Thailand, Turkey, Ukraine, and Vietnam, and one regional investment plan in the Middle East and North Africa (MENA) covering Algeria, Egypt, Jordan, Morocco, and Tunisia. Projects include support for wind energy, urban public transportation systems, solar water heaters, smart-grid development, and concentrating solar thermal power programs, among others (see *Table 3* for more detailed descriptions of the national investment plans).

Governance

The CTF is implemented through a partnership of the multilateral development banks (MDBs) and governed by representatives from the donor and recipient countries. The role of governance for the CTF is to approve investment plans, programming, and the allocation of financial resources; and to provide guidance, performance evaluation, and reporting. It is further tasked with ensuring that the strategic orientation of the CTF is guided by the principles of the UNFCCC. The organizational structure of the CTF is equally balanced between donor and developing countries. All decisions are made by consensus. Other international organizations, the private sector, and civil society representatives are included as observers. All observer roles are "active," allowing them to take the floor to make interventions, propose agenda items, and recommend experts. Observers do not vote during consensus decisions. The governance structure includes the following:

- The CTF Trust Fund Committee, which oversees and decides on the operations and activities of the CTF and includes (1) eight representatives from contributor countries; (2) eight representatives from eligible recipient countries; (3) a representative from the project recipient country (during deliberations on the investment plan, program, or project); (4) a representative of the World Bank; and (5) a representative for the other MDBs.
- The MDBs Committee, which facilitates collaboration, coordination, and the exchange of information, knowledge, and experience among MDBs partners.
- The Partnership Forum, which supports civil society engagement and includes representatives of donor and eligible recipient countries, MDBs, U.N. and U.N. agencies, Global Environment Facility (GEF), UNFCCC, Adaptation Fund, bilateral development agencies, NGOs, indigenous peoples, private sector entities, and technical experts.
- The Administrative Unit, which supports the work of the CIFs, is housed in the World Bank's Washington, DC, offices.
- A Trustee (the World Bank), which holds in trust, as the legal owner and administrator, the funds, assets, and receipts that constitute the Trust Fund, pursuant to the terms entered into with the contributors.

Funding

Since September 2008, 14 donor countries have pledged over US$7.6 billion (in historical value) to finance the two CIF trust funds. The total amount pledged by the nine contributing countries to the CTF has been US$5.154 billion (in historical value)[10] as of March 31, 2013 (see *Table 2* for pledges and contributions; *Table 1* for U.S. Budget Authority). The funds are to be disbursed as grants, concessional loans, loan guarantees, and other risk management instruments. Endorsed funding by the CIFs also serves to leverage co-financing from additional sources, including the private sector, multilateral financial institutions, recipient governments, state-owned enterprises, and carbon finance.

Table 2. Total Pledges and Contributions to the Clean Technology Fund As of March 31, 2013 (USD millions)

Donor	Contribution Type[a]	Amount Pledged (historical value)[b]	Amount Pledged (current value)[c]	Receipts (current value)[d]
Australia	Grant	$84	$86	$86
Canada	Loan	$193	$199	$199
France	Loan	$300	$268	$268
Germany	Loan	$739	$615	$615
Japan	Grant	$1,000	$1,114	$1,114
Spain	Capital	$118	$109	$109
Sweden	Grant	$92	$80	$80
United Kingdom	Capital	$1,135	$973	$973
United States[e]	Grant	$1,492	$1,492	$714
Total		$5,154	$4,937	$4,158

Source: The CIFs website at http://www.climateinvestmentfunds.org/

[a] Donor contribution types include grants, loans, and equity, and describe in broad terms the general requirements stipulated by the donors on their contributed funds. The U.S. government has historically contributed grant financing for reasons that include ease, ODA accounting practices, and flexible capital reflow provisions.

[b] Represents pledges valued on the basis of exchange rates as of September 25, 2008, the CIF official pledging date.

[c] Valued on the basis of exchange rates as of December 31, 2012.

[d] Valued on the basis of exchange rates as of December 31, 2012.

[e] The total U.S. pledge to the CIFs remains at $2 billion. Contributions across funds are extrapolated from current allocations.

Program Areas

The CTF is based on country and regional investment plans that aim to support climate-friendly technologies. Investment plans are undertaken jointly by the recipients, the MDBs, other development partners, private industry, and civil society to build upon existing national strategies and demonstrate how the CTF can be complementary to the country's overall developmental activities. The CTF supports investment plans that are cost-effective and implementation-ready, can be scaled up quickly to impact development, and have the potential for significant greenhouse gas emission reductions. To receive CTF funding, a country must be eligible for official development assistance (ODA) and have an active MDB program.

The majority of CTF funding supports programs that help shape demand side markets for technology diffusion. The fund's criteria for lending allow for all renewable and energy efficiency initiatives, as well as large-scale hydroelectric power plants, natural gas plants, some forms of biofuels, power plant refits, and ultra-supercritical coal plants.[11] Funds are commonly targeted to support a variety of investment activities, including (1) direct purchase of technological goods and services; (2) direct investment into government infrastructure for transport or transmission modernization; (3) seed funds for financial intermediaries to incentivize clean technology lending; and (4) investment support and risk mitigation strategies for private sector entry into the market. In short, the CTF attempts to address the additional costs contained in lower-carbon energy investment such that it becomes a viable option to conventional fossil-fuel power generation. *Table 3* outlines the endorsed investment plans as of March 31, 2013.

Table 3. Clean Technology Fund Investment Plans (In USD millions)

Date of Endorsement / Revision	Country	Direct CTF Funding / Co-financing	Investment Plan
January 2009 / November 2012	Egypt	$300 / $1,817	Wind power; Urban transport (natural gas buses and a subway); Transmission upgrades.
January 2009	Mexico	$500 / $6,624	Energy efficiency (appliance & lighting); Urban transport (rapid bus); Wind power.
January 2009 / Novermber 2012	Turkey	$250 / $2,378	Renewable energy and energy efficiency; Smart-grid technology.
October 2009 / October 2011	Morocco	$150 / $2,470	Energy sector privatization; Energy conservation; Urban transport.
October 2009	South Africa	$500 / $3,147	Concentrated solar power; Wind power; Solar water heaters; Energy efficiency.
December 2009	Middle East / N. Africa region	$750 / $4,391	Concentrated solar power; Transmission and distribution infrastructure.
December 2009 / February 2012	Thailand	$170 / $216	Renewable energy and efficiency; Urban transport (bus system).
December 2009 / August 2012	Philippines	$250 / $2,095	Solar power; Transmission infrastructure; Demand side management; Sustainable transport strategy.
December 2009 /	Vietnam	$250 / $3,978	Renewable energy and industrial

Date of Endorsement / Revision	Country	Direct CTF Funding / Co-financing	Investment Plan
June 2011			energy efficiency; Urban transport (rail system); Initial capitalization of funds;
			Transmission infrastructure.
March 2010	Colombia	$150 / $2,340	Sustainable transport program; Public/private sector energy efficiency program.
March 2010	Indonesia	$400 / $2,630	Large-scale geothermal power; Biomass and other renewable energy.
March 2010	Kazakhstan	$200 / $719	Hydro and wind power; Public sector transport fuel switch; District heating; Energy efficiency.
March 2010	Ukraine	$350 / $2,012	Wind, hydro, biomass; Residential and government energy efficiency; District heating; Smartgrid technology.
November 2010	Nigeria	$250 / $510	Transport sector structure; Clean and renewable energy development; Energy efficiency; Financial sector reform.
November 2011	India	$775 / $4,069	Energy efficiency; Large-scale solar.
May 2012	Chile	$200 / $610	Concentrated solar power; Large-scale solar PV; energy efficiency and small-scale self-supply.
November 2012	Turkey-Stage 2	$140 / $397	Energy efficiency, energy finance.

Source: CTF committee meeting documents and national Investment plans, available at the CIFs website.

The Strategic Climate Fund (SCF)

Overview

Some governments and civil society organizations are concerned that climate change may exacerbate poverty situations and reverse economic gains in the developing world through the possibility of temperature increases, rising sea levels, droughts, changes in rainfall patterns, heightened disease patterns, and the lack of drinkable water. They believe that resources may be necessary to help low-income countries manage a response. Responses to climate change are likely to entail both mitigation efforts (i.e., slowing, then reducing greenhouse gas emissions) and adaptation efforts (i.e., managing the effects of short- and long-term climate outcomes). The SCF aims to help developing countries prepare for climate change by promoting low-carbon, climate-resilient development. Three targeted programs provide grants and concessional loans to pilot new approaches aimed at specific challenges:[12]

- The Pilot Program for Climate Resilience (PPCR) supports ways to integrate climate risk and resilience into the development strategies of low-income countries. Funds can be used to provide technical assistance to help with capacity building, policy reform, and sector investment.
- The Forest Investment Program (FIP) provides financing to countries to help them prepare for and participate in programs that aim to reduce deforestation. Funds can be used for managing forests and for educating indigenous and local communities about forest policies.
- The Scaling Up Renewable Energy Program in Low Income Countries (SREP) helps low-income countries adopt renewable energy solutions to aid in the development of their power generation sector. Funds can be used to provide policy support, technical assistance, financial management, and sector investment.

Governance

The SCF is implemented through a partnership of the multilateral development banks (MDBs) and governed by representatives from the donor and recipient countries. The governance and decision-making structure is similar to the CTF, but specifically includes the following:

- The SCF Trust Fund Committee, which oversees and decides on the operations and activities of SCF and includes (1) eight representatives from contributor countries; (2) eight representatives from eligible recipient countries; (3) a representative of the World Bank; and (4) a representative for the other MDBs.
- An SCF subcommittee for each of the targeted programs, which includes up to six representatives from contributor countries to the SCF Program, a matching number of representatives from eligible recipient countries, and such other representatives designated by the SCF Trust Fund.

- The MDBs Committee, which facilitates collaboration, coordination, and the exchange of information, knowledge, and experience among the MDBs partners.
- The Partnership Forum, which supports civil society engagement and includes representatives of donor and eligible recipient countries, MDBs, U.N. and U.N. agencies, GEF, UNFCCC, Adaptation Fund, bilateral development agencies, NGOs, indigenous peoples, private sector entities, and technical experts.
- The Administrative Unit, which supports the work of the CIFs, is housed in the World Bank's Washington, DC, offices.
- A Trustee (the World Bank), which holds in trust, as the legal owner and administrator, the funds, assets, and receipts that constitute the Trust Fund, pursuant to the terms entered into with the contributors.

Funding

Since September 2008, 14 donor countries have pledged over US$7.6 billion (in historical value) to finance the two CIF trust funds. The total amount pledged by 13 countries to the SCF has been US$2.413 billion (in historical value)[13] as of March 31, 2013 (see *Table 4* for pledges and contributions; *Table 1* for U.S. Budget Authority). The funds are to be disbursed as grants, concessional loans, loan guarantees, and other risk management instruments.

Program Areas

The programming of the SCF is newer than that of the CTF, having launched no earlier than January 2009. However, each of the funds has begun endorsing investment plans. As of the April 30, 2013 meeting of the Joint CTF and SCF Trust Fund Committees, the status of each fund was reported as follows:[14]

- *The Pilot Program for Climate Resilience.* The PPCR became operational in January 2009. The program provides funding to the countries in two phases: (1) a technical assistance phase, which includes looking at how countries' development plans can be made more climate-resilient and deciding upon the types of investments countries could make; and (2) an implementation phase, which includes the dispersal of grants of up to $1.5 million with the option of additional loans to implement programs. The PPCR Sub-

Committee has endorsed 20 investment plans for $1,034.4 million in PPCR funding. This includes plans for 18 countries (Bangladesh, Bolivia, Cambodia, Dominica, Grenada, Haiti, Jamaica, Mozambique, Nepal, Niger, Papua New Guinea, Samoa, St. Lucia, St. Vincent and the Grenadines, Tajikistan, Tonga, Yemen, and Zambia) and two regional programs (Caribbean Regional Program and the Pacific Regional Program). The plans are expected to leverage an additional $1.47 billion in co-financing. *Table 5* outlines the endorsed investment plans as of March 31, 2013.

- *The Forest Investment Program.* The FIP became operational in February 2010. The FIP Sub-Committee has endorsed seven investment plans (Brazil, Burkina Faso, DR Congo, Ghana, Indonesia, Lao PDR, and Mexico) for $370 million in FIP funding. These plans are projected to leverage an additional $993 million in co-financing. An investment plan for Peru is expected to be endorsed in November 2013.
- *The Scaling Up Renewable Energy Program in Low Income Countries.* The SREP become operational in December 2009. The SREP Sub-Committee has endorsed six investment plans (Ethiopia, Honduras, Kenya, Maldives, Mali, and Nepal) for $240 million in SREP funding. The plans are expected to leverage an additional $1.74 billion in co-financing. Investment plans for Liberia and Tanzania are expected for endorsement by November 2013.

Table 4. Total Pledges and Contributions to the Strategic Climate Fund As of March 31, 2013 (USD millions)

Donor	Contribution Type[a]	Amount Pledged (historical value)[b]	Amount Pledged (current value)[c]	Receipts (current value)[d]
Australia	Grant	$72	$79	$79
Canada	Grant	$97	$84	$84
Denmark	Grant	$47	$45	$45
Germany	Grant	$88	$78	$78
Japan	Grant	$200	$218	$218
Korea	Grant	$6	$6	$6
Netherlands	Grant	$76	$76	$76
Norway	Grant	$241	$229	$229
Spain	Grant /Capital	$34	$30	$30
Sweden	Grant	$42	$41	$41
Switzerland	Grant	$26	$26	$26

Donor	Contribution Type[a]	Amount Pledged (historical value)[b]	Amount Pledged (current value)[c]	Receipts (current value)[d]
United Kingdom	Capital	$976	$838	$838
United States[e]	Grant	$508	$508	$200
Total		$2,413	$2,257	$1,950

Source: The CIFs website at http://www.climateinvestmentfunds.org/

[a] Donor contribution types include grants, loans, and equity, and describe in broad terms the general requirements stipulated by the donors on their contributed funds. The U.S. government has historically contributed grant financing for reasons that include ease, ODA accounting practices, and flexible capital reflow provisions.

[b] Represents pledges valued on the basis of exchange rates as of September 25, 2008, the CIF official pledging date.

[c] Valued on the basis of exchange rates as of December 31, 2012.

[d] Valued on the basis of exchange rates as of December 31, 2012.

[e] The total U.S. pledge to the CIFs remains at $2 billion. Contributions across funds are extrapolated from current allocations.

Table 5. Pilot Program for Climate Resilience Investment Plans (In USD millions)

Date of Endorsement	Country	Direct PPCR Funding/ Co-financing	Investment Plan
November 2010	Bangladesh	$110 / $566	Enabling environment; infrastructure; coastal zone management; agriculture and landscape management.
November 2010	Niger	$110 / $2	Agriculture and landscape management; climate information systems and disaster risk management; water resources management.
November 2010	Tajikistan	$58 / $84	Climate information systems and disaster risk management; enabling environment; agriculture and landscape management; water resources management; infrastructure.

Table 5. (Continued)

Date of Endorsement	Country	Direct PPCR Funding/ Co-financing	Investment Plan
April 2011	Caribbean - Grenada	$25 / $13	Climate information systems and disaster risk management; agriculture and landscape management.
April 2011	Caribbean - St.	$15 / $12	Climate information systems and disaster risk management.
	Vincent and the Grenadines		
April 2011	Pacific - Samoa	$30 / $15	Infrastructure; coastal zone management.
June 2011	Cambodia	$91 / $340	Infrastructure, enabling environment; water resources management; agriculture and landscape management.
June 2011	Mozambique	$91 / $190	Agriculture and landscape management; enabling environment; water resources management; infrastructure; urban development.
June 2011	Nepal	$91 / $20	Enabling environment; climate information systems and disaster risk management; agriculture and landscape management; water resources management.
June 2011	Zambia	$91 / $116	Agriculture and landscape management; enabling environment.
June 2011	Caribbean - St. Lucia	$27 / $15	Climate information systems and disaster risk management.
November 2011	Bolivyesia	$91 / $49	Water resources management.
November 2011	Caribbean - Jamaica	$30 / $18	Enabling environment; climate information

Date of Endorsement	Country	Direct PPCR Funding/ Co-financing	Investment Plan
			systems and disaster risk management; water resources management.
April 2012	Caribbean -	$21 / $17	Climate information systems and disaster risk management.
	Dominica		
April 2012	Yemen	$58 / $5	Climate information systems and disaster risk management; coastal zone management.
April 2012	Caribbean - Regional Track	$11 / $11	Enabling environment.
April 2012	Pacific - Tonga	$20 / $0	Enabling environment.
April 2012	Pacific - Regional Track	$11 / $0	Enabling environment; coastal zone management.
November 2012	Pacific - Papua	$30 / $0	Agriculture and landscape management.
	New Guinea		
May 2013	Caribbean - Haiti	$25 / $110	Agriculture and landscape management; climate information systems; infrastructure; urban development.

Source: PPCR committee meeting documents and national Investment plans, available at the CIFs website.

CURRENT ISSUES

Each year, billions of dollars in environmental aid flow from developed country governments— including the United States—to developing ones.[15] While the efficiency and the effectiveness of these programs are of concern to donor country governments, a full analysis of the purposes, intents, results, and consequences behind these financial flows has yet to be conducted.[16] International relations, comparative politics, and developmental economics can often collide with global environmental agendas.

Critics contend that the existing system has had limited impact in addressing major environmental concerns—specifically climate change and tropical deforestation—and has been unsuccessful in delivering global transformational change. A desire to achieve more immediate impacts has led to a restructuring of the MDBs' role in environmental finance and the introduction of many new bilateral and multilateral funding initiatives. The CIFs grew out of these concerns.

The effectiveness of the CIFs depends on how the trust funds address their programmatic issues, build upon their national investment plans, react to recent developments in the financial landscape, and respond to emerging opportunities.

The following section investigates some of the current challenges facing the CIFs and summarizes some of the responses initiated by the funds.

Innovations by the CIFs

Since their inception, the CIFs have attempted to provide innovative approaches to global environmental issues and have introduced several processes to address the limitations of previous environmental finance.[17] These innovations include, but are not limited to, the following:

- *Programmatic Design.* While the CIFs still aim to scale up existing practices and fund activities at the project level, they also were created to serve as laboratories for new financing schemes and vehicles for developing sustainable strategies. Funding strives to target the potential for large-scale transformation and to attain global environmental benefits. Stakeholders seek to share knowledge gained and inspire the use of best practice. As such, multinational orregional investment plans that support global development goals, energy security, industrial growth, diversification, and regional integration (e.g., the M.E.N.A. plan) best exemplify the CIFs' programmatic approach.
- *Country-led Process.* Beyond a simple project-by-project approach, the purpose of the CIFs is to bolster the efforts of countries' official adaptation plans and their actions toward low-carbon, climate-resilient development. The country-led approach aims to integrate funding into the country-owned development strategies consistent with the Paris Declaration.[18]
- *Innovative Governance and Stakeholder Engagement.* In an effort to attain transparency and accountability, the governing structure of the CIFs is equally balanced between donor and developing countries. All

decisions are taken by consensus, with no provision for voting. If a consensus is not possible, the proposal is postponed or withdrawn. Representatives from other international organizations, the private sector, and civil society are included as observers. All observer roles are "active," allowing them to take the floor to make interventions, propose agenda items, and recommend experts.

Issues in Support of the Multilateral Development Banks (MDBs) and Multilateral Assistance

The choice of financial mechanism and its administration is an important element to environmental finance. The differences among multilateral or bilateral assistance, grant or lending institutions, regional or global organizations, etc., all play a role in the structure of assistance. The decision to employ the MDBs as trustees for the CIFs has both advantages and disadvantages. Historically, the MDBs have provided financial assistance to developing countries, typically in the form of loans and grants, for investment projects and organizational capacity.[19] Donor country support for the MDBs—including U.S. support—has assisted efforts to promote institutions, strengthen financial systems, undertake large infrastructure and social welfare projects, and develop property rights and rules of law. Through increased global integration, the aim of the MDBs has been to bolster economic growth, poverty alleviation, and resource allocation (including greater access to electricity) in developing countries while simultaneously building new markets for developed countries' exports and jobs. In 2008, at the urging of some donor countries,[20] a strategy to address climate change was added to the MDBs' development agenda. The "Strategic Framework on Development and Climate Change"[21] analyzed the risks of climate change to economic development and served as a basis for integrating mitigation and adaptation planning into national development plans. Donor countries see several advantages to financing climate programs through the institutional structure of the MDBs. These advantages include, but are not limited to, the following:

- *Commitment to Private Sector Development.* Many donor countries—including the United States—believe that climate-friendly economic growth can be led by the private sector through such efforts as improving access to financial markets, building the capacity of entrepreneurs, and providing training to civilian society. One aim of the MDBs is to help foster private sector development by leveraging donor funds into highly effective co-financing arrangements. Historically, the U.S. Administration has supported these efforts. In a March 25, 2010, hearing before the House Appropriations Subcommittee on State, Foreign Operations, and Related Programs, the Treasury Department went on record as stating that the United States invests in the MDBs because "they help generate new engines of growth that benefit the U.S. economy and the global economy as a whole."

- *Economies of Scale, Coordination, and Co-financing.* Proponents of the MDBs argue that multilateral assistance can solve problems of scale and efficiency by providing specialized expertise while lowering administration and coordination costs. Similarly, more competitive procurement rules, attractive cost-sharing opportunities, and the ability to leverage co-financing from other public and private organizations allow the MDBs to play a catalytic role in mobilizing financial aid.[22] At the March 25, 2010, hearing noted above, the Treasury Department stated that the MDBs "provide strong, effective and highly leveraged means to advance global prosperity.... For every dollar the United States contributes to paid-in capital for the World Bank, six dollars of additional capital is generated by other donors. And, for every dollar we invest in the World Bank, $26 worth of aid is delivered."[23]
- *Responsiveness to Donors.* The Treasury Department has similarly stated that the United States invests in the MDBs because they "promot[e] core American interests and values." This arrangement is due primarily to the structure and organization of the banks. MDBs' governance is weighted on the basis of the cumulative financial contributions and commitments by the donor countries, and thus, while a single trust fund, like the CIFs, may be designed to balance equally the roles of developed and developing countries, the MDBs are designed to give greater weight to the major donors. The United States retains the most influence on World Bank matters, with a 14.97% voting share and the ability to veto major policy decisions. It is followed by Japan in second place, Germany in third, and France and the United Kingdom tied for fourth. The only developing or emerging country with as much voting interest is China, at fifth, with 3.21%.[24] With a governing structure that requires one representative from the World Bank, as trustee, and one representative from the

group of remaining MDBs, as well as eight representatives from participating donor countries, the overall governance structure of the CIFs has remained responsive to donor interests.

- Possession of Fiduciary Standards. Both current and past U.S. Administrations have argued that the World Bank has the proper internal safeguards to oversee large amounts of financing. As reported by the Department of Treasury, "the World Bank is an attractive trustee [for environmental funds] precisely because of its strong fiduciary standards and its extensive capacity to uphold them."25
- Possession of Institutional Expertise, Information, and Credibility Provisions. Proponents of the MDBs claim that multilateral agencies offer larger and better trained staffs with greater technical expertise. They state that large infrastructure investment, particularly in innovative technologies and methods, requires professionals who are experienced in identifying and facilitating access to technology, sharing risks associated with commercialization, and improving institutional capacity. Beyond institutional knowledge, multilaterals also collect, interpret, and disseminate costly information on a global scale and provide credibility controls for both recipient and donor governments.

Issues of Concern for Developing Countries and NGOs

While advantages exist to financing climate programs through the institutional structure of the MDBs, concerns also persist. A variety of recipient countries and nongovernmental organizations (NGOs)[26] have highlighted a number of issues, including, but not limited to, the following:

- *Coordination with Other Funds.* Proponents argue that the fundamental principle of the CIFs is coordination at the country-level among interested stakeholders, including other developmental partners. However, some observers believe that the CIFs have created a parallel structure for financing climate change efforts outside both bilateral and the ongoing multilateral framework for climate change negotiations. They are concerned that without harmonization between the CIFs and the other sources of environmental finance (e.g., funds managed by the U.N., the Global Environment Facility, and bilateral sources), overlaps, redundancies, competing views, and lack of synergy may affect climate priorities, funding processes, and qualifying criteria.

- *Potential to Prejudice U.N. Climate Provisions.* Some commentators and several governments have expressed concerns that the establishment of the CIFs as trust funds in the MDBs may prejudice the outcomes of the international negotiations on climate finance within the framework of the United Nations. Many developing countries have expressly stated that they do not consider funds contributed to the CIFs as meeting U.N. Annex I obligations. Furthermore, the design of the CIFs includes a "sunset clause" stating that the CIFs "will take necessary steps to conclude its operations once a new [UNFCCC] financial architecture is effective." The nature of these steps has yet to be determined. Further, additional contributions to the CIFs beyond the initial 2008 pledges (e.g., by Canada ($193 million), United Kingdom (£375 million), Denmark ($8 million), Germany ($12 million), Norway ($19 million), Sweden ($25 million), and Switzerland ($6 million)) may complicate these negotiations.
- *Potential for Additionality.* The UNFCCC provides that developed country signatories to the Convention "provide new and additional financial resources to meet the agreed full costs incurred by developing country Parties" in their efforts at mitigation and adaptation.[27] Some observers fear that the design of the CIFs establishes a parallel process for climate financing that does not result in new and additional resources. They are concerned that significant portions of the aid budgets of donors may be diverted into the CIFs and counted as part of their annual ODA commitments.
- *Lack of Polluter Responsibility.* Some commentators claim that the provision of loans as a financial instrument to eligible developing countries contradicts the internationally agreed principle of "polluter pays" as stated in the Rio Declaration. Some argue that the repayment of a loan, notwithstanding the degree of concessionality,[28] burdens a developing country with self-paying for a problem (climate change) that was caused by others (i.e., developed countries). They believe this burden may affect the country's ability to generate resources for growth.
- *Commercial Influence.* While advantages exist in prioritizing market-based solutions to dealing with the problems of climate, some groups express concern that the private sector may be unduly driven by

commercial interests at the expense of social or environmental safeguards.[29] Concern also exists that a dependence on market mechanisms as a source of climate financing may be inadequate and inconsistent for meeting the financial needs of developing countries charged with the responsibility of both implementing climate change commitments and mediating the social, economic, and environmental dislocations brought on by climate change.

- *Energy and Environmental Policy at the Banks.* Many observers claim that the history of the World Bank's energy and infrastructure lending undermines its credibility as an institution committed to combating the impacts of climate change. Environmental NGOs have often highlighted the inconsistencies between the Bank's rhetoric on climate change and its operational policies and practices. They emphasize that while the Bank has increased financing for renewable energy and energy efficiency in recent years, its fossil fuel lending still accounts for 56% of the energy sector share for fiscal years 2008 to 2010 (compared to 15% for renewable energy, 20% for energy efficiency, and 9% for large hydropower).[30] The controversy is compounded by the Bank's inability to reach a consensus on the definition of "clean energy technology," retaining provisions for ultra-supercritical coal-fired power generation in its environmental strategies.[31] Recent guidance from the U.S. Administration regarding the World Bank's engagement with coal-fired power generation in developing countries similarly leaves the definition open, stating that projects "could include more carbon efficient fossil fuel generation" in their portfolio.[32] While observers generally agree that funding from the CIFs is unlikely to be used in coal-fired power generation projects, most agree that continued investment by the World Bank in fossil fuel energy and infrastructure may have several unintended effects, including (1) counteracting any gains made with the Bank's renewable portfolio, (2) directing resources toward large-scale power generation for industrial use rather than energy access and poverty reduction in poor urban and rural communities, and (3) drawing the Bank's professional and technical staff away from a concentration on energy efficiency and renewable energy activities to remain involved with fossil fuels.[33]

End Notes

[1] For a more detailed discussion on various sources and mechanisms of financial assistance for climate change activities, see CRS Report R41808, International Climate Change Financing: Needs, Sources, and Delivery Methods, by Richard K. Lattanzio and Jane A. Leggett.

[2] The group of multilateral development banks referred to in this report includes the World Bank Group (WBG), African Development Bank (AfDB), Asian Development Bank (ADB), European Bank for Reconstruction and Development (EBRD), and Inter-American Development Bank Group (IDB).

[3] For more substantive analysis of foreign aid and congressional roles, see CRS Report R40213, Foreign Aid: An Introduction to U.S. Programs and Policy, by Curt Tarnoff and Marian Leonardo Lawson; and CRS Report R41170, Multilateral Development Banks: Overview and Issues for Congress, by Rebecca M. Nelson.

[4] As summarized by the International Institute for Sustainable Development, at http://www.iisd. ca/download/pdf/sd/ ymbvol172num2e.pdf.

5 See the World Bank website for additional information at http://siteresources.worldbank. org/NEWS/Resources/ Climate_Change_Results_Brief_4-12-10.pdf.

[6] Henry Paulson, Alistair Darling, and Fukushiro Nukaga, "Financial bridge from dirty to clean," Financial Times, February 7, 2008.

[7] For a full description of purpose and programs, see the CIFs website at http://www.climate investmentfunds.org/cif/.

[8] Valued on the basis of exchange rates as of December 31, 2012, the last recorded Trustee Report for the CIFs, at https://www.climateinvestmentfunds.org/cif/funding-basics.

[9] U.S. Department of State, FY 2014 Executive Budget Summary - Function 150 and Other International Programs, at http://www.state.gov/f/budget/

[10] As of December 31, 2012, current value of the pledges was USD equivalent $4.937 billion.

[11] No coal-fired power plants have been proposed or approved at this time. Hydroelectric power generation is currently included in the Ukraine proposal.

[12] Description of SCF overview and governance from CIFs, Annual Report 2009, on the CIFs website at http://www.climateinvestmentfunds.org/cif/sites/climateinvestmentfunds. org /files/cif_annual_report_final_021810.pdf.

[13] As of December 31, 2012, current value of the pledges was USD equivalent $2.257 billion.

[14] See SCP Committee document, "SCF/TFC.10/3/Rev.1, Progress report on SCF targeted programs," at https://www.climateinvestmentfunds.org/cif/workingdocuments/11015.

[15] The Organisation for Economic Co-operation and Development (OECD) maintains information on Member countries' Official Development Assistance. Current data (accessed April 15, 2011) reflect that all OECD DAC Member countries contributed, on average, a total of $2,283 million per annum over the period 2005-2009 to multisectoral environmental protection assistance (in 2010 US$), and that the United States contributed, on average, $285 million per annum over the same period to multi-sectoral environmental protection assistance (in 2010 US$). See OECD StatExtracts database at http://stats.oecd. org/Index.aspx?DataSetCode=ODA_DONOR#.

[16] This report does not aim to unpack the full range of discussions on environmental and developmental assistance. For a discussion on international development assistance in general, see CRS Report R40213, Foreign Aid: An Introduction to U.S. Programs and Policy, by Curt Tarnoff and Marian Leonardo Lawson. An overview and analysis of the history of environmental financing can be found in a number of source materials including

recent book length studies by Inge Kaul and Pedro Conceição, The New Public Finance: Responding to Global Challenges, New York: Oxford University Press, 2006; and Robert L. Hicks, Bradley C. Parks, J. Timmons Roberts, and Michael J. Tierney, Greening Aid?: Understanding the Environmental Impact of Development Assistance, New York: Oxford University Press, 2008.

[17] For further discussion regarding the limitations of past mechanisms for global environmental finance, see the section on institutional challenges in CRS Report R41165, International Environmental Financing: The Global Environment Facility (GEF), by Richard K. Lattanzio.

[18] The 2005 Paris Declaration, endorsed by over 100 countries, aims to increase harmonization, alignment, and management of aid for results with a set of actions and indicators that can be monitored. See http://www.oecd.org/ dataoecd/11/41/34428351.pdf.

[19] For a fuller discussion on the structure and the role of the MDB system, refer to CRS Report R41170, Multilateral Development Banks: Overview and Issues for Congress, by Rebecca M. Nelson.

[20] Including the United States. See the negotiations at the 2005 G8 Gleneagles Summit at http://www.g7.utoronto.ca/ summit/2005gleneagles/.

[21] See http://siteresources.worldbank.org/EXTCC/Resources/407863-1219339233881/DCCSFT echnicalReport.pdf.

[22] Sources of additional funds most often include other MDBs and multilateral financial institutions, the recipient governments, state-owned enterprises, and carbon finance, as well as the private sector.

[23] See testimony at http://appropriations.house.gov/images/stories/pdf/sfo/Secretary_Geithner. 3.25.10.pdf.

[24] As reported on the World Bank website, "Voting Powers," at http://go.worldbank. org/VKVDQDUC10. These figures are for country voting shares at the International Bank for Reconstruction and Development (IBRD). Shares for the IFC, IDA, and MIGA may vary.

[25] As reported in Climatewire, "Eskom fallout spurs new opposition to World Bank's role in climate funding," May 24, 2010.

[26] There are many published critiques on the environmental agenda of the MDBs. Of specific relevance for CIFs, see, for example, Celine Tan, "No Additionality, New Conditionality: A Critique of the World Bank's Proposed Climate Investment Funds," TWN, 2008, at http://www.twnside.org.sg/bangkok.briefings.htm; Smita Nakhooda, "Catalyzing Low-carbon Development?" WRI, 2009, at http://pdf.wri.org/working_papers/ development_ clean_technology_fund.pdf; and Heike Mainhardt-Gibbs, et al., "Fuelling Contradictions: The World Bank's Energy Lending and Climate Change," Bretton Woods Project, CRBM & URGEWALD, 2010, at http://www.brettonwoodsproject.org/art-566198.

[27] See UNFCCC, Article 4:3, at http://unfccc.int/essential_background/convention/ background/ items/1362.php.

[28] "Concessional" or "soft" loans are loans extended on terms substantially more generous than market loans. The concessionality is achieved either through interest rates below those available on the market or by extended grace periods, or a combination of these.

[29] This concern has been levied against the Bank's brokering of carbon purchases through its Prototype Carbon Fund. See Bank Information Center, et al., "How the World Bank's Energy Framework Sells the Climate and Poor People Short," 2006, at http://www. bicusa.org/en/Article.2954.aspx.

[30] See Bank Information Center, "World Bank Group Energy Sector Financing Update," November 2010, at http://www.bicusa.org/en/Document.102339.pdf.

[31] Ultra-supercritical coal-fired power generation is defined as "new pulverised coal combustion systems ... [that] operate at increasingly higher temperatures and pressures and therefore achieve higher efficiencies than conventional PCC units and significant CO2 reductions. Supercritical steam cycle technology has been used for decades and is becoming the system of choice for new commercial coal-fired plants in many countries." See World Coal Institute website at http://www.worldcoal.org/coal-the-environment/ coal-use-the-environment/improving-efficiencies/.

[32] See U.S. Treasury memorandum at http://www.treasury.gov/resource-center/international/development-banks/Pages/ guidance.aspx.

[33] For discussion of further debate on this issue, see the World Bank's issue brief on "Energy," available at http://go.worldbank.org/E084GP3GQ0; and, as one example, Heike Mainhardt-Gibbs, et al., "Fuelling Contradictions: The World Bank's Energy Lending and Climate Change," the Bretton Woods Project, CRBM & URGEWALD, 2010, at http://www.brettonwoodsproject.org/art-566198.

In: Climate and Environmental Protection
Editor: Jackson M. Garcia
ISBN: 978-1-63117-960-0

Chapter 2

INTERNATIONAL CLIMATE CHANGE FINANCING: THE GREEN CLIMATE FUND (GCF)*

Richard K. Lattanzio

SUMMARY

Over the past several decades, the United States has delivered financial and technical assistance for climate change activities in the developing world through a variety of bilateral and multilateral programs. The United States and other industrialized countries committed to such assistance through the United Nations Framework Convention on Climate Change (UNFCCC, Treaty Number: 102-38, 1992), the Copenhagen Accord (2009), and the UNFCCC Cancun Agreements (2010), wherein the higher-income countries pledged jointly up to $30 billion of "fast start" climate financing for lower-income countries for the period 2010-2012, and a goal of mobilizing jointly $100 billion annually by 2020. The Cancun Agreements also proposed that the pledged funds are to be new, additional to previous flows, adequate, predictable, and sustained, and are to come from a wide variety of sources, both public and private, bilateral and multilateral, including alternative sources of finance.

* This is an edited, reformatted and augmented version of a Congressional Research Service publication, CRS Report for Congress R41889, prepared for Members and Committees of Congress, from www.crs.gov, dated April 16, 2013.

One potential mechanism for mobilizing a share of the proposed international climate financing is the UNFCCC Green Climate Fund (GCF), proposed in the Cancun Agreements and accepted by Parties during the December 2011 conference in Durban, South Africa. The fund aims to assist developing countries in their efforts to combat climate change through the provision of grants and other concessional financing for mitigation and adaptation projects, programs, policies, and activities. The GCF is to be capitalized by contributions from donor countries and other sources, including both innovative mechanisms and the private sector. Currently, the GCF complements many of the existing multilateral climate change funds (e.g., the Global Environment Facility, the Climate Investment Funds, and the Adaptation Fund); however, as the official financial mechanism of the UNFCCC, some Parties believe that it may eventually replace or subsume the other funds. While many Parties expect capitalization and operation of the GCF to begin shortly after the November 2013 conference in Warsaw, Poland, many issues remain to be clarified, and some involve long-standing and contentious debate. They include what role the CGF would play in providing sustained finance at scale, how it would fit into the existing development assistance and climate financing architecture, how it would be capitalized, and how it would allocate and deliver assistance efficiently and effectively to developing countries.

The U.S. Congress—through its role in authorizations, appropriations, and oversight—would have significant input on U.S. participation in the GCF. Congress regularly determines and gives guidance to the allocation of foreign aid between bilateral and multilateral assistance as well as among the variety of multilateral mechanisms. In the past, Congress has raised concerns regarding the cost, purpose, direction, efficiency, and effectiveness of the UNFCCC and existing international institutions of climate financing. Potential authorizations and appropriations for the GCF would rest with several committees, including the U.S. House of Representatives Committees on Foreign Affairs (various subcommittees); Financial Services (Subcommittee on International Monetary Policy and Trade); and Appropriations (Subcommittee on State, Foreign Operations, and Related Programs); and the U.S. Senate Committees on Foreign Relations (Subcommittee on International Development and Foreign Assistance, Economic Affairs, and International Environmental Protection); and Appropriations (Subcommittee on State, Foreign Operations, and Related Programs). As of April 2013, the U.S. Administration—through its State, Foreign Operations, and Related Programs 150 account—has made no specific budget request for appropriated funds to be contributed to the GCF.

INTRODUCTION[1]

Many voices, domestic and international, have called upon the United States and other industrialized countries to increase foreign assistance to lower- and middle-income countries to address climate change.[2] Proponents maintain that such assistance could help promote climate-friendly and high-growth economic development in these countries, while simultaneously protecting the more vulnerable nations from the effects of a changing climate. For their part, most, if not all, lower-income countries have stated that their success at combating climate change depends critically on receipt of international financial support. They argue that mitigating climate change pollutants, adapting to the effects of climate change, and building climate resilience into their development agendas incur costs above and beyond their normal economic growth trajectories. These costs are particularly challenging to nations that have scant resources compared to industrialized countries, do not recognize themselves as the historical sources of climate pollution, and consider alleviating poverty as their first priority.

The Green Climate Fund (GCF) is an international financial institution connected to the United Nations Framework Convention on Climate Change (UNFCCC).[3] The GCF was proposed by Parties to the UNFCCC during the 2009 Conference of Parties (COP) in Copenhagen, Denmark, and its design was agreed to during the 2011 COP in Durban, South Africa. The fund aims to assist developing countries in their efforts to combat climate change through the provision of grants and other concessional financing for mitigation and adaptation[4] projects, programs, policies, and activities. The GCF is to be capitalized by contributions from donor countries and other sources, including both innovative mechanisms and the private sector. Currently, the GCF complements many of the existing multilateral climate change funds (e.g., the Global Environment Facility, the Climate Investment Funds, and the Adaptation Fund), however, as the official financial mechanism of the UNFCCC, some Parties believe that it may eventually replace or subsume the other funds. Expectations by many countries, specifically developing countries, are that the GCF becomes very large (i.e., in the range of several tens of billions to over $100 billion annually) and serves as the predominant institution for climate change assistance in the developing world.[5] These countries believe that the agreement to establish the GCF has been a key success in the recent international negotiations.

But others caution that ambitious steps need to be taken to ensure that the fund is implemented and operated correctly in order to achieve an adequate buy-in by donor countries of its effectiveness and by recipient countries of its legitimacy.

The GCF is currently under implementation. While the governing Board, a host city, and the basic design of the fund have been agreed to by Parties, many structural aspects of the GCF have yet to be determined. These include legal arrangements between the fund and the COP, setting of operational procedures, establishment of a Secretariat, and the final determination of the trustee.

Prospective dates to complete these various arrangements have been laid out over the course of 2013, and many Parties expect capitalization and operation of the GCF to begin shortly after the November 2013 conference in Warsaw, Poland. There remain issues that need to be clarified, however, and some involve long-standing and contentious debate. They include

- what role the CGF would play in providing sustained finance at adequate levels;
- how it would fit into the existing development assistance and climate financing architecture;
- how it would be capitalized; and
- how it would allocate and deliver assistance efficiently and effectively to developing countries.

The U.S. Congress—through its role in authorizations, appropriations, and oversight—would have significant input on U.S. participation in the fund. Congressional input on the GCF may include

- whether and when to participate in the fund;
- whether and how much to contribute to the fund, and with what source or sources of finance;
- whether fund contributions would carry specific guidance in distribution and use;
- how contributions to the fund would relate to other U.S. bilateral, multilateral, and private sector climate change assistance; and
- whether and when to consent to negotiated treaty obligations, if submitted.

BACKGROUND ON THE UNITED NATIONS FRAMEWORK CONVENTION ON CLIMATE CHANGE

The UNFCCC was the first formal international agreement to acknowledge and address human-driven climate change. The U.S. Senate provided its advice and consent to the Convention's ratification in 1992, the same year it was concluded.[6] For the United States, the UNFCCC entered into force in 1994. As of January 2011, 194 governments are Parties. As a framework convention, the UNFCCC provides a structure for international consideration of climate change but does not contain detailed obligations for achieving particular climate-related objectives in each Party's territory. It recognizes that climate change is a "common concern to humankind," and, accordingly, requires parties to (1) gather and share information on GHG emissions, national policies, and best practices; (2) launch national strategies for addressing GHG emissions and adapting to expected impacts; and (3) cooperate in preparing for the impacts of climate change. The UNFCCC did not set binding targets for GHG emissions; however, it did commit the higher-income Parties (i.e., those listed in Annex II of the Convention)[7] to provide unspecified amounts of financial assistance to help lower-income countries meet the broad, qualitative obligations common to all Parties.[8]

As the treaty entered into force and the UNFCCC Conference of the Parties (COP) met for the first time in 1995, the Parties agreed that achieving the objective of the UNFCCC would require new and stronger GHG commitments. As a first step toward meeting this objective, the 1997 Kyoto Protocol was drafted and entered into force with a stated aim to reduce the net GHG emissions of industrialized country Parties (Annex I Parties) to 5.2% below 1990 levels in the period of 2008 to 2012. The United States signed the Kyoto Protocol in December 1997. However, at the time, opposition in Congress was strong.[9] The Kyoto Protocol was not submitted to the Senate by President Clinton or by his successor, President George W. Bush. Thus, the United States is not a Party to the Protocol.

In 2007, UNFCCC Parties reconvened negotiations for further commitments beyond the Kyoto Protocol, and agreed to negotiate a suite of agreements that included new GHG mitigation targets for Annex I Parties, "nationally appropriate mitigation actions" for non-Annex I Parties, and other commitments for the post-2012 period. The mandates (referred to as the Bali Action Plan) specify that the products of negotiation should be ready by the end of 2009. Due perhaps to high expectations, as well as continued divergence among Parties on some key issues, the 2009 COP in Copenhagen, Denmark, did not produce a legally binding treaty, but a short, non-legally binding political document called the Copenhagen Accord.[10]

The Copenhagen Accord is a policy document drafted by leaders of about two dozen countries in the final hours of the 2009 COP, and subsequently acknowledged by 114 countries. The Accord sits in sharp contrast to the Kyoto Protocol, as its bottom-up and nationally appropriate model differs greatly from the top-down implementation of the Protocol. Provisions in the Accord include voluntary GHG mitigation efforts by all Parties, adaptation and forestry actions, technology transfer mechanisms, and transparency and reporting standards, as well as financial provisions by developed country Parties. The Accord also establishes the "Green Climate Fund" (GCF) to serve as the operating entity of the financial mechanism of the Convention.

Many of the elements of the Copenhagen Accord, the Bali Action Plan, and the UNFCCC were adopted officially at the 2010 COP in Cancun, which yielded several decisions collectively called the Cancun Agreements.[11] The establishment of the GCF—as well as some other financial arrangements mentioned in the Copenhagen Accord—was a central aspect of the negotiations, and was entered into the negotiating text of the UNFCCC's Ad Hoc Working Group on Longterm Cooperative Action under the Convention (AWG-LCA).[12] Climate finance provisions in the Cancun negotiating text (CP.16) include the following:

- *Fast Start Financing.* The agreement puts forth a collective commitment by developed country Parties (not specified in the text) to provide new and additional resources approaching $30 billion for the period 2010–2012 to address the needs of developing countries (the allocation, or "burden-sharing," among countries is not specified in the text) (CP.16§95).
- *2020 Pledge.* The agreement takes note of the pledge by developed country Parties (not specified in the text) to achieve a goal of mobilizing jointly $100 billion per year by 2020 to address the needs of developing countries (the allocation, or "burden-sharing," among countries is not specified in the text) (CP.16§98).
- *Sources.* The agreement outlines that the pledged assistance is to be scaled-up, new and additional, predicable and adequate, and that it may come from a wide variety of sources, including public and private, bilateral, multilateral, and alternative (CP.16§§97, 99).[13]

- *Balanced Package.* The agreement recognizes that the financial pledges are offered in the context of continued negotiations toward a balanced package of commitments by all Parties that includes, among other items, meaningful actions on mitigation[14] and transparency[15] (CP.16§98).
- *Green Climate Fund.* The agreement opens the way for the establishment of the GCF, to be designated as an operating entity of the financial mechanism of the UNFCCC, accountable to and under the guidance of the COP, to support projects, programs, policies, and other activities in developing country Parties (CP.16§102).
- *Transitional Committee.* The agreement stipulates the formation of a Transitional Committee to design the fund, comprising 40 members, with 15 members from developed country Parties and 25 members from developing country Parties, with experience and skills in the areas of finance and climate change, in accordance with given Terms of Reference (CP.16§§109–110).

The Durban Platform and the Green Climate Fund

The basic design of the GCF, as recommended by the Transitional Committee, was adopted officially at the 2011 COP in Durban, South Africa, which yielded several decisions collectively called the "Durban Platform."[16] The design of the GCF was a central aspect of the negotiations, and the approved mandates were entered into the negotiating text of the UNFCCC's Ad Hoc Working Group on Long-term Cooperative Action under the Convention (AWG-LCA). Decisions on the GCF in the Durban negotiating text (CP.17)[17] include the following:

- *Status.* The agreement designates the GCF as "the operating entity of the Financial Mechanism of the Convention" to be "accountable to and function under the guidance of the Conference of Parties to support projects, programmes, policies and other activities in developing country Parties" (CP.17§3).
- *Governance.* The agreement sets forth the composition of a Board, to have 24 members, composed of an equal number from developing and developed country Parties, with representation from relevant United

Nations groupings including Small Island States (SIDS) and Least Developed Countries (LDC) (CP.17§A9). The decision invites Parties to submit their nominations for Board membership by March 31, 2012 (CP.17§10). Functions of the Board are to include designing operations, establishing funding windows, approving funding, selecting implementing agencies, defining an accreditation process for implementing agencies, developing fiduciary standards and environmental and social safeguards, and building a framework for the monitoring and evaluation of performance.

- *Host Country.* The agreement begins the process of selecting a host country for the GCF, asking Parties to submit expressions of interest by April 15, 2012, and requiring a final decision for endorsement by the 18th session of the COP (CP.17§§12-13).
- *Management.* The agreement tasks the Board with establishing an independent Secretariat to execute the day-to-day operations of the fund, to be in place no later than by the 19th session of the COP (CP.17§§15,19).
- *Trustee.* The agreement asks the Board to open a transparent and competitive bidding process for the selection of a trustee, either to replace or continue the services of the World Bank, as interim trustee (CP.17§16).
- *Board Meetings.* The agreement also authorizes the Board to set up an interim Secretariat immediately with the goal of convening the first Board meeting by April 30, 2012. The first two board meetings are to be hosted by Switzerland and South Korea (CP.17§24).

THE DOHA GATEWAY AND THE GREEN CLIMATE FUND

The GCF Board formed in 2012 and began a series of meetings to decide on recommendations to bring before the UNFCCC.[18] Parties to the 2012 COP in Doha, Qatar, endorsed the consensus decision of the GCF Board to select Songdo, Incheon, Republic of Korea as the host of the GCF. The GCF Board and the Republic of Korea were asked to conclude the legal and administrative arrangements for hosting the fund, and to ensure that juridical personality and legal capacity are conferred to the GCF.[19] Further, Parties recognized that the Governing Instrument[20] for the GCF formed the basis for the arrangements between the COP and the GCF, and asked the Board to develop the final arrangements for such an agreement.[21]

At the November 2013 conference in Poland, the GCF Board is scheduled to report on the progress of the implementation of Decision 3/CP.17, which included the following requests:

- to develop a transparent no-objection procedure to be conducted through national designated authorities;
- to balance the allocation of the resources of the GCF between adaptation and mitigation activities;
- to secure funding for the GCF to facilitate its expeditious operationalization, and to establish the necessary policies and procedures to enable an early and adequate replenishment process;
- to establish the independent secretariat of the GCF;
- to select the trustee of the GCF through an open, transparent, and competitive bidding process in a timely manner to ensure that there is no discontinuity in trustee services;
- to initiate a process to collaborate with the Adaptation Committee and the Technology Executive Committee, as well as other relevant thematic bodies under the Convention, to define linkages between the Fund and these bodies.

Furthermore, Parties decided to provide initial guidance to the GCF at COP 19, and requested the GCF Board to expeditiously implement its 2013 work plan, with a view to making the GCF operational as soon as possible.

Outstanding Challenges for the Fund

Observers[22] have noted that the GCF still confronts many challenges in design, scope, governance, and implementation that must be decided either by the Board or by the Parties to the UNFCCC. Several of the most significant challenges left outstanding or unanswered by the Durban negotiating text include the following:

Relationship of the Fund to the Convention

As currently conceived, the GCF is intended to operate at arm's length from the UNFCCC, with an independent Board, Trustee, and Secretariat. The Durban negotiating text states that the GCF is to be "accountable to and function under the guidance of the Conference of Parties" (CP.17§A4) (i.e., similar in legal structure to the Global Environment Facility), as opposed to "accountable to and function under the guidance *and authority* of the Conference of Parties" (i.e., similar in legal structure to the Adaptation Fund). While subtle, the distinction carries import, and negotiators from China and the Group of 77[23] have—for the moment at least—kept the latter structure in conversation in an effort to ensure representation by all Parties of the UNFCCC. The majority of developed country Parties, however, oppose the Adaptation Fund model as inefficient and overly politicized for two key reasons: (1) the COP would have direct authority over the selection and release of all Board and Secretariat members, and (2) the COP would have final approval over all rules and guidelines proposed by the Board. Keeping the fund independent from the COP has been a key negotiating point for the United States. Given the current language of the negotiating text, and the proposed design of the Board to carry equal representation between developed and developing country Parties, this issue may already be resolved.

The World Bank As Trustee

The role of the World Bank in the GCF has been, and continues to be, controversial. The Durban negotiating text confirms the World Bank as the interim trustee, subject to review after three years of fund operation (CP.17§A26). Most believe that once established, a subsequent shift in institutional arrangement is doubtful. Many developing countries hold the World Bank in a negative light, believing it to be non-transparent, overly bureaucratic, and reflecting solely the interests of higher-income countries, which command greater decision-making power by virtue of their greater financial contributions. Additionally, some Parties see the potential for conflicts of interest during the implementation phase of the fund, since the Bank (1) already operates a portfolio of Climate Investment Funds that might compete against the GCF for potential donor country contributions, and (2) has been asked to serve as support staff to aid in designing the operational procedures, project selection criteria, performance standards, and safeguard measures for the new fund. Despite these concerns, some Parties remain unconvinced that an adequate substitute exists, claiming that no other extant institution could undertake the proposed financial administration and fiduciary standards with the same level of confidence from the donors.

Mobilization of Funds

The Cancun negotiating text is silent on sources, with no proposal for how finance would flow into the fund. The text simply "takes note" of "relevant reports on the financing needs and options for mobilization of resources to address the needs of developing country Parties with regard to climate change adaptation and mitigation, including the report of the High-level Advisory Group on Climate Change Finance [AGF]" (CP.16§101). The Durban negotiating text elaborates little on the sources of funding beyond two short statements: (1) the fund "would receive financial inputs from developed country Parties to the Convention," and (2) the fund "may also receive financial inputs from a variety of other sources, public and private, including alternative sources" (CP.17§§A29-A30). Most see the faint mention of sources as unsurprising. The 2010 report by the AGF concluded that the goal of mobilizing adequate and predictable climate finance to developing countries on the order of $100 billion annually would be "challenging but feasible."[24] Further, while it is acknowledged that adequate international finance would likely require a range of sources (including public finance, development bank instruments, carbon markets, and private capital), little unity exists among COP Parties as to the balance between public and private sources, developed and developing country participation in international carbon markets or tax schemes, and the political feasibility of other large-scale fund mobilizations. Several other multilateral fora have taken up the issue of climate finance sourcing, including the G-20 and the Major Economies Meeting.[25] However, the means by which the issue may be resolved during the fund's implementation is unclear.

Operational Modalities

The Durban negotiating text outlines several design aspects regarding the operation of the fund, including "complementarity, eligibility, structure, access modalities, and financial instruments" (CP.17§§A31-A56). Each category engenders debate among Parties, and the negotiating text leaves a number of issues open for consideration during implementation.

- *Complementarity.* While little has been decided regarding the eventual size and scope of the GCF, its formation is being viewed by many as a means through which to simplify the complex network of multilateral and bilateral funding mechanisms that currently provide climate change assistance to developing countries. Many early proponents of a global fund had envisioned that such an institution would play the role of a "fund of funds," or an "umbrella," under which to collect both the resources and the comparative advantages of the other mechanisms. As it currently stands, the Durban negotiating text gives little indication that such an ambition is to be pursued by the GCF. It states, instead, that the fund would "operate in the context of appropriate arrangements between itself and other existing funds under the Convention, and between itself and other funds, entities, and channels of climate change financing outside the Fund" (CP.17§A33). Nevertheless, the fate of the other funds has been called into question by the establishment of the GCF. At present, the Adaptation Fund is the sanctioned U.N. mechanism in support of climate change assistance for adaptation actions. The Global Environment Facility is the sanctioned UNFCCC financial mechanism in support of mitigation actions. The World Bank's Climate Investment Funds are designed to sunset in 2012 at the presumed commencement of the new UNFCCC mechanism. It is possible that the eventual scope of the GCF may overshadow and/or replace these funds.[26] Conversely, it is also possible that the CGF may prove inadequate to existing arrangements in the eyes of potential donors.
- *Eligibility.* The Durban negotiating text states that "all developing country Parties to the Convention are eligible to receive resources from the Fund" (CP.17§A35). Presumably this characterization would include middle-income countries like Brazil, India, South Africa, and China. The United States is on record as objecting to this arrangement.[27]
- *Structure.* While the Copenhagen Accord specifies that the GCF would support activities related to "mitigation including REDD-plus,[28] adaptation, capacity building, technology development and transfer" (CP.16§10), the Cancun negotiating text dropped such references, opting instead to state that the GCF would use "thematic

funding windows" (CP.16§102). The Durban text further unsettles the structure of the fund by stating that the GCF would "initially have windows for adaptation and mitigation"; but would likewise "ensure adequate resources for capacity-building and technology development and trade" as well as "consider the need for additional windows" (CP.17§§A37-A39). Further, the Board is tasked with "balancing" the allocation of resources between adaptation and mitigation (CP.17§A50). With present funding by existing financial institutions decidedly tilted toward mitigation actions,[29] there is likely to be a strong expectation—by developing countries as well as certain civil society organizations—that adaptation actions receive a significant portion of support from the CGF. Currently, no allocation formula has been provided, nor has a definition of "balance."

- *Access.* Consideration of how countries would access funds from the GCF, and which agencies and organizations would be allowed to acquire funds to implement projects, remains an ongoing issue of debate. Currently, almost all multilateral financial assistance for climate change activities in developing countries is channeled through third-party implementing agencies (e.g., U.N. agencies, multilateral development banks, major nongovernmental organizations).[30] The Durban text invites international entities to provide services for the GCF; however, it emphasizes "direct access" modalities as a way to enhance recipient country ownership in the process (CP.17§§A45-48). Direct access has become a prominent, new arrangement in climate finance delivery, allowing the recipient country to access financial resources directly from the fund, and/or allowing it to assign an implementing agency of its own choosing. This operational freedom has been a rallying point for many developing country Parties. The modality is also supported by many developed country Parties as a means to secure broader competition and greater country ownership. Nevertheless, implementation of direct access arrangements may prove to be slow and difficult, because they would likely require the same stringent level of fiduciary standards, competitive procurement practices, and environmental and social safeguards demanded of existing third-party implementing agencies. The Durban text places the burden of developing "an accreditation process for all implementing entities" on the Board (CP.17§A49).
- *Instruments.* As for the choice of instruments, the Durban negotiating text states that financing would be provided "in the form of grants and

concessional lending, and through other modalities, instruments or facilities as may be approved by the Board" (CP.17§A54). Observers stress that climate finance can take a variety of forms; however, debate consistently arises between donor and recipient countries as to the appropriateness of debt-based instruments (i.e., loans) for humanitarian aid. While the general presumption is that climate finance in support of adaptation actions in developing countries should be provided on grant terms, this is less customary with regard to mitigation actions. Thus, many see it as important for the GCF to secure a good match between the type of finance and the object of financing, retaining sufficient funds to provide grants when necessary, as influenced by both the country and the project profile.

RELATIONSHIP OF THE FUND TO OTHER U.S. CLIMATE FINANCE COMMITMENTS

At present, there have been no official funds pledged to the GCF. Further, there have been no pronouncements regarding redirection of current or future bilateral or multilateral financial assistance into the fund. Nor have the eventual size, scope, or sources of financial contributions been decided in the negotiating text or voiced officially by the UNFCCC, Parties to the COP, or members of the Transitional Committee. The only specific financial allocation tied directly to the GCF in the Cancun negotiating text is "that a significant share of new multilateral funding for adaptation *should* flow through the Green Climate Fund" (CP.16§100, italics added).[31] This statement does not define "new" or "significant"; does not identify specific contributors or recipients; and does not assign specific amounts, allocations, or changes to bilateral aid, mitigation assistance, or current levels of funding. The Durban negotiating text elaborates little on the sources of funding beyond two short statements: (1) the fund "would receive financial inputs from developed country Parties to the Convention," and (2) the fund "may also receive financial inputs from a variety of other sources, public and private, including alternative sources" (CP.17§§A29-30).

In regard to the GCF's relationship to other climate finance commitments by the United States:

- *UNFCCC Fast Start Financing Pledges.* The collective pledge by developed country Parties to provide new and additional resources approaching $30 billion for the period 2010-2012 is *not* tied to the GCF. In fact, the GCF is unlikely to be fully operational before the end of 2012. Fast start funds most likely would continue as current bilateral and multilateral contributions made through authorized appropriations by donor country governments.
- *UNFCCC 2020 Pledges.* The collective pledge by developed country Parties to the goal of mobilizing jointly $100 billion per year by 2020 is *not* tied directly to the GCF. The Cancun negotiating text makes clear that "funds provided to developing country Parties may come from a wide variety of sources, public and private, bilateral and multilateral, including alternative sources" (CP.16§99). When operational, the GCF would be one of many possible public and multilateral sources. While any financial assistance that is channeled through the GCF would likely be considered a part of the $100 billion goal, the entirety of the $100 billion goal is not expected to be provided solely by the GCF, and no estimation of the GCF's presumed share has been suggested officially. Many Parties, as well as the AGF report, have suggested that development bank instruments, carbon markets, and—especially—private capital would be critical to mobilizing assistance at the level pledged.
- *Bilateral Aid.* The GCF would not necessarily interfere with current or proposed bilateral climate change assistance to developing countries. The GCF would be another multilateral mechanism for climate change assistance that would exist alongside bilateral activities, much the way that the Global Environment Facility and the Climate Investment Funds currently do. U.S. allocations between and among bilateral and multilateral assistance channels would continue through authorized congressional appropriations.
- *Other Multilateral Aid.* The GCF Board has been tasked with determining the complementarity of the GCF with respect to other U.N. multilateral mechanisms. Thus, the negotiations may produce some alteration in the landscape of the multilateral choices provided by the UNFCCC. Development bank mechanisms such as the Climate Investment Funds may or may not be reevaluated by their governing Boards in light of the final implementation of the GCF. Presumably, choices would remain available to donor countries. U.S. allocations

among multilateral assistance channels would remain based on congressional guidance and would continue through authorized congressional appropriations.

Issues for Congress

Members of Congress hold mixed views about the value of international financial assistance to address climate change. While some Members are convinced that human-induced climate change is a high-priority risk that must be addressed through federal actions and international cooperation, others are not as convinced. Some are wary, as well, of international processes that could impose costs on the United States, redirect funds from domestic budget priorities, undermine national sovereignty, or lead to competitive advantages for other countries. Regardless of current views, the United States is a Party to the UNFCCC and has certain obligations under the treaty. The executive branch continues negotiations and implementation of the UNFCCC obligations, while committees of Congress engage in oversight (from home and at the international meetings), providing input to the executive branch formally and informally, and deciding program authorities and appropriations for these activities.

As Congress considers potential authorization and/or appropriations for the GCF, it may raise concerns regarding the cost, purpose, direction, efficiency, and effectiveness of the UNFCCC and existing international financial institutions. These concerns may be weighed against the design characteristics of the GCF in an effort to assess its potential performance. Congress may then be required to determine the allocation of funds between bilateral and multilateral climate change assistance as well as among the variety of multilateral mechanisms. Congress may also wish to gauge and give guidance to the new fund's relationship with domestic industries and private sector investment, as well as the spillover effects of U.S. participation on technological innovation, humanitarian efforts, national security, and international leadership. Potential authorizations and appropriations for the GCF rest with several committees, including the U.S. House of Representatives Committees on Foreign Affairs (various subcommittees); Financial Services (Subcommittee on International Monetary Policy and Trade); and Appropriations (Subcommittee on State, Foreign Operations, and Related Programs); and the U.S. Senate Committees on Foreign Relations (Subcommittee on International Development and Foreign Assistance,

Economic Affairs, and International Environmental Protection); and Appropriations (Subcommittee on State, Foreign Operations, and Related Programs).

Additional issues for Congress concerning the climate negotiations in general, and the GCF in particular, may include the means to establish a more desirable form of agreement (or lack thereof); the compatibility of any international agreement with U.S. domestic policies and laws; the adequacy of appropriations and fiscal incentives to achieve any commitments under the agreement; and any requirements for potential ratification and implementing legislation, should a formal treaty emerge from the negotiations.

End Notes

[1] This report assumes a general understanding of climate change science, policy, financing, and international negotiations. For further background on climate change science and policy, see CRS Report RL34266, Climate Change: Science Highlights, by Jane A. Leggett, and CRS Report RL34513, Climate Change: Current Issues and Policy Tools, by Jane A. Leggett; for further background on international climate change financing, see CRS Report R41808, International Climate Change Financing: Needs, Sources, and Delivery Methods, by Richard K. Lattanzio and Jane A. Leggett; and for further background on the international climate change negotiations, see CRS Report R40001, A U.S.-Centric Chronology of the International Climate Change Negotiations, by Jane A. Leggett.

[2] Most industrialized countries currently deliver some financial and technical assistance through a variety of bilateral development programs and multilateral financial institutions. "Bilateral" assistance involves direct transfers from one country to another; "multilateral" assistance is distributed through international organizations and agencies such as the United Nations and the World Bank Group.

[3] The UNFCCC and its processes are discussed in greater detail in the next section, "Background on the United Nations Framework Convention on Climate Change."

[4] "Mitigation activities" refer to actions taken to reduce or reverse the forces that contribute to global climate change (examples include transitioning to a low-emissions energy supply; capturing the opportunities in energy efficiency improvements in buildings, transportation, and industry; reducing deforestation and improving sustainable forest management to better serve as GHG emissions sinks; and employing more low-emissions and sustainable agriculture practices). "Adaptation activities" refer to adjustments made in natural or human systems in response to actual or expected climate change and its effects (examples include employing climate-resistant crop varieties, improving irrigation systems, integrating sustainable land management into agricultural planning, protecting water resources, managing coastal zones, designing infrastructure for extreme weather or for sea-level rise, and improving public health services).

[5] Many developing countries have stated their support for a large and centralized fund for climate change assistance to be housed at the UNFCCC, capitalized primarily by public contributions, and funded at or near the level of the estimated costs for climate change activities in the developing world. While cost estimates vary greatly depending upon the

analysis (see "Cost Estimates" section in CRS Report R41808, International Climate Change Financing: Needs, Sources, and Delivery Methods, op cit.), many countries have acknowledged the annual $100 billion figure as a target. To put this figure in context, in FY2010 the United States provided $1.3 billion for international climate change assistance, split almost equally between bilateral and multilateral programs. Further, the FY2010 U.S. budget authority for all foreign operations programs, both bilateral and multilateral, at the Departments of State, Treasury, and the Agency for International Development, totaled $32.8 billion, or approximately 3% of all FY2010 U.S. discretionary spending. Beyond the existing multilateral funds, total global contributions to international climate change assistance have been difficult to track due to the lack of international standards. Transparency of financial contributions and standardized accounting procedures are included in the package of agreements currently under negotiation by Parties to the UNFCCC.

[6] United Nations Framework Convention on Climate Change, Treaty Number: 102-38, October 7, 1992, the resolution of advice and consent to ratification agreed to in the Senate by Division Vote.

[7] UNFCCC Annex I Parties include the industrialized countries that were members of the OECD (Organization for Economic Cooperation and Development) in 1992, plus countries with economies in transition (the EIT Parties), including the Russian Federation, the Baltic States, and several Central and Eastern European States. Annex II Parties consist of the OECD members of Annex I, but not the EIT Parties.

[8] For more information on the UNFCCC and U.S. participation in international climate treaties, see CRS Report R41175, International Agreements on Climate Change: Selected Legal Questions.

[9] See the Byrd-Hagel Resolution (S.Res. 98) in July 1997, wherein the Senate expressed its opposition (95-0 vote) to the terms of the Berlin Mandate (the 1995 UNFCCC COP agreement that led to the adoption of the Kyoto Protocol) by stating that the United States should not sign any treaty that does not include specific, scheduled commitments of non-Annex I Parties in the same compliance period as Annex I Parties, or that might seriously harm the U.S. economy.

[10] See UNFCCC Decision 2/CP.15, Copenhagen Accord, FCCC/CP/2009/11/Add.1, at http://unfccc.int/resource/docs/ 2009/cop15/eng/11a01.pdf.

[11] See UNFCCC Decision 1/CP.16, The Cancun Agreements: Outcome of the work of the Ad Hoc Working Group on Long-term Cooperative Action under the Convention, FCCC/CP/2010/7/Add.1 at http://unfccc.int/resource/docs/2010/ cop16/eng/07a01. pdf# page=2.

[12] The Ad Hoc Working Group on Long-term Cooperative Action under the Convention (AWG-LCA) is the negotiating track of the COP that serves "to enable the full, effective and sustained implementation of the Convention through long-term cooperative action, now, up to, and beyond 2012." This is in contrast to the Ad Hoc Working Group on Further Commitments for Annex I Parties under the Kyoto Protocol (AWG-KP) (for which the United States is not a Party, but an observer). Both Working Groups are tasked with producing negotiating texts with the aim of moving the texts to an agreement among Parties to the UNFCCC COP.

[13] The United Nations convened an advisory panel in 2010 to investigate issues regarding the sourcing of climate finance. The report by the U.N. High Level Advisory Group on Climate Change Financing was released in November 2010 and can be found at http://www.un.org/ wcm/content/site/climatechange/pages/financeadvisorygroup. For a discussion of the

various proposed public, private, and alternative sources, as well as the issues revolving around the mobilization of international climate finance, see also CRS Report R41808, International Climate Change Financing: Needs, Sources, and Delivery Methods, by Richard K. Lattanzio and Jane A. Leggett.

[14] "Mitigation" commitments refer to the formalized pledges taken by developed country Parties, or the nationally appropriate measures being taken by developing country Parties, to reduce GHG emissions.

[15] "Transparency" commitments refer to the negotiated provisions whereby all major economy Parties (including the large emerging economies) must report on the progress they are making in meeting their mitigation commitments (targets and actions), and all Parties providing financial assistance must report their contributions through commonly accepted formats.

[16] See UNFCCC Decisions accepted by COP 17, at http://unfccc.int/2860.php.

[17] See UNFCCC Decision -/CP.17, Green Climate Fund—Report of the Transitional Committee and Annex, at http://unfccc.int/files/meetings/durban_nov_2011/ decisions/application/ pdf/cop17_gcf.pdf. Text appearing in the "Annex" is referenced in the textual notes with a letter "A" before the paragraph number.

[18] For a list of GCF Board members, see the fund's website, http://gcfund.net/ board/composition.html. Currently, Matthew Kotchen, Deputy Assistant Secretary, Office of Environment and Energy, Department of the Treasury, serves on the Board for the United States.

[19] See Decision 6/CP.18, Report of the Green Climate Fund to the Conference of the Parties and guidance to the Green Climate Fund, at http://unfccc.int/resource/docs/2012/cop18 /eng/08a01.pdf.

[20] The "Governing Instrument for the Green Climate Fund" was approved by the Conference of the Parties to the UNFCCC at its 17th session on December 11, 2011 in Durban, South Africa, and is annexed to Decision 3/CP.17 presented in UNFCCC document FCCC/CP/2011/9/Add.1 (see http://unfccc.int/resource/docs/2011/cop17/eng/ 09a01.pdf.

[21] See Decision 7/CP.18, Arrangements between the Conference of the Parties and the Green Climate Fund, at http://unfccc.int/resource/docs/2012/cop18/eng/08a01.pdf.

[22] Many economic and environmental civil society organizations have watched and reported on the UNFCCC climate negotiations in general, and the GCF negotiations in particular. For more analyses on the design challenges outlined in this report, among others, as well as on other proposed fund models, see, for example, Neil Bird, Jessica Brown, and Liane Schalatek, Design Challenges for the Green Climate Fund, Climate Finance Policy Brief No. 4, Heinrich Böll Foundation North America and Overseas Development Institute, January 2011; Nigel Purvis and Andrew Stevenson, "Climate Negotiations and International Finance," Resources, Winter/Spring 2011, No. 177, pp. 16-18; and Benito Müller, Time to Roll Up the Sleeves—Even Higher!: Longer-term Climate Finance after Cancun, Oxford Energy and Environment Brief, Climate Strategies and the Oxford Institute for Energy Studies, January 2011.

[23] The Group of 77 is an official U.N. negotiating group composed of developing countries. Founded in 1964 in the context of the U.N. Conference on Trade and Development (UNCTAD), it now has over 130 members.

[24] See AGF Report, op cit.

[25] The "G-20" refers to the Group of Twenty: a forum for finance ministers and central bank governors established in 1999 to bring together systemically important industrialized and developing economies to discuss key issues in the global economy. The "Major Economies

Forum" refers to a forum of 17 major developed and developing economies established in 2009 to facilitate dialogue on energy and climate issues.

[26] Recent funding levels for the above-mentioned multilateral funds are as follows: Global Environment Facility, $3.5 billion pledged for the period 2011-2014; Climate Investment Funds, $6.1 billion pledged for the period 2009-2012; Adaptation Fund, based on a formula of 2% of the Certified Emission Reduction units issued for projects of the Clean Development Mechanism of the UNFCCC Kyoto Protocol.

[27] U.S. Congress, House Committee on Foreign Affairs, Subcommittee on Oversight and Investigations, U.N. Climate Talks and Power Politics, 112th Cong., 1st sess., May 25, 2011, S.Hrg. 112-22, p. 20, wherein Todd Stern, U.S. Special Envoy for Climate Change, in responding to questions, stated that "after I arrived in Copenhagen in 2009, I did my first press conference. And I was asked about funding for China. And I said I didn't really anticipate that U.S. funds, which are limited in any event, would be most wisely spent going to China."

[28] "REDD-plus," or "Reducing Emissions from Deforestation and Forest Degradation," activities refer to mitigation-relevant activities in support of forestry and sustainable land management.

[29] Some of the reasons donor countries more readily provide financial assistance for mitigation projects as opposed to adaptation projects may include (1) mitigation actions serve to benefit the global environment, whereas adaptation actions often only provide benefits at the local level; (2) mitigation actions in the form of large-scale infrastructure projects associated with low-carbon technological development often are prioritized by donor countries because they support—and are incentivized by—global investment and the private sector, whereas local adaptation projects often find private sector mobilization more difficult to facilitate; and (3) mitigation actions in the form of large-scale infrastructure projects often are easier to monitor, verify, assess, and evaluate compared to smaller-scale, local adaptation projects.

[30] The only current exception to this practice is the U.N. Adaptation Fund, which has begun to accredit national agencies in recipient countries as official implementing agencies for fund disbursement. Council Members for the Global Environment Facility have begun discussions on a similar practice. This practice is referred to as "direct access."

[31] It has been suggested by some Parties that the inclusion of this statement may be the result of Parties' experiences with the U.N. Adaptation Fund, which serves as a reminder of the challenge to raise funds directly from countries' national budgetary contributions and the reluctance of such countries to commit a significant share of these funds to UNFCCC-sponsored adaptation projects.

In: Climate and Environmental Protection ISBN: 978-1-63117-960-0
Editor: Jackson M. Garcia

Chapter 3

INTERNATIONAL ENVIRONMENTAL FINANCING: THE GLOBAL ENVIRONMENT FACILITY (GEF)*

Richard K. Lattanzio

SUMMARY

This report provides an overview of one of the oldest international financial institutions for the environment—the Global Environment Facility (GEF)—and analyzes its structure, funding, and objectives in light of the many challenges within the contemporary landscape of global environmental finance.

GEF is an independent and international financial organization that provides grants, promotes cooperation, and fosters actions in developing countries to protect the global environment. Established in 1991, it unites 182 member governments and partners with international institutions, nongovernmental organizations, and the private sector to assist developing countries with environmental projects related to six areas: biodiversity, climate change, international waters, the ozone layer, land degradation, and persistent organic pollutants. GEF receives funding from multiple donor countries—including the United States—and provides grants to cover the additional or "incremental" costs associated with

* This is an edited, reformatted and augmented version of a Congressional Research Service publication, CRS Report for Congress R41165, prepared for Members and Committees of Congress, from www.crs.gov, dated June 3, 2013.

transforming a project with national benefits into one with global environmental benefits. In this way, GEF funding is structured to "supplement" base project funding and provide for the environmental components in national development agendas. GEF partners with several international agencies, including the International Bank for Reconstruction and Development, the United Nations Development Program (UNDP), and the United Nations Environment Program (UNEP), among others, and is the primary fund administrator for four Rio (Earth Summit) Conventions, including the Convention on Biological Diversity (CBD), the United Nations Framework Convention on Climate Change (UNFCCC), the Stockholm Convention on Persistent Organic Pollutants (POPs), and the United Nations Convention to Combat Desertification (UNCCD). GEF also establishes operational guidance for international waters and ozone activities, the latter consistent with the Montreal Protocol on Substances that Deplete the Ozone Layer and its amendments. Since its inception, GEF has allocated $11.5 billion—supplemented by more than $57 billion in cofinancing—for more than 3,200 projects in over 165 countries.

GEF is one mechanism in a larger network of international programs designed to address the global environment. Accordingly, its effectiveness depends on how the fund addresses programmatic issues, builds upon national investment plans, reacts to recent developments in the financial landscape, and responds to emerging opportunities. Critics contend that the existing system has had limited impact in addressing major environmental concerns—specifically climate change and tropical deforestation—and has been unsuccessful in delivering global transformational change. A desire to achieve more immediate impacts has led to a restructuring of the Multilateral Development Banks' (MDBs') role in environmental finance and the introduction of many new bilateral and multilateral funding initiatives. The future of GEF remains in the hands of the donor countries, including the United States, which can choose to broaden the mandate and/or strengthen its institutional arrangements or reduce and replace it by other bilateral or multilateral funding mechanisms.

THE GLOBAL ENVIRONMENT FACILITY

The Global Environment Facility (GEF) is an independent and international financial organization that provides grants, promotes cooperation, and fosters actions in developing countries to protect the global environment. Established in 1991, it unites 182 member governments and partners with international institutions, nongovernmental organizations, and the private sector to assist developing countries with environmental projects related to six areas: biodiversity, climate change, international waters, the ozone layer, land degradation, and persistent organic pollutants (POPs). GEF receives funding from multiple donor countries[1]— including the United States—and provides grants to cover the additional or "incremental" costs associated with transforming a project with national benefits into one with global environmental benefits (e.g., choosing solar energy technology over coal technology meets the same national development goal of power generation but is more costly, excluding long-term environmental externalities; GEF grants aim to cover the difference or "increment" between a less costly, more polluting option and a costlier, more environmentally sound option). In this way, GEF funding is structured to "supplement" base project funding and provide for the environmental components in national development agendas. Since its inception, GEF has allocated $11.5 billion— supplemented by more than $57 billion in co-financing—for more than 3,200 projects in over 165 countries.[2]

Background

The idea for a Global Environment Facility was proposed in a September 1989 meeting of the joint International Bank for Reconstruction and Development (the World Bank)—International Monetary Fund Development Committee after recommendation by a World Resources Institute report commissioned by the United Nations.[3] The fund was established in 1991 as a pilot program within the World Bank, and many observers saw it as the beginning of an important shift in multilateral policy toward incorporating environmental concerns into development assistance. GEF, however, quickly ran into some operational challenges. These included (1) problems with communication among the implementing agencies (i.e., among the World Bank economists, the United Nations Development Program engineers, and the United Nations Environment Program environmentalists), (2) problems with differing agendas among the donor countries (i.e., between environmental idealism in Europe and economic pragmatism in the United States and United Kingdom), and (3) problems with differing perspectives among developing countries (i.e., between an emphasis on economic growth or environmental initiatives).

Initially, GEF had been opposed by developing countries who believed that a program established and controlled by higher-income donor countries under the framework of the Multilateral Development Banks was not in their best interest. They remained committed to a governing structure and a cooperative partnership founded on a U.N.-style majority-based decision. After three years of debate, GEF was restructured in 1994 to address many of its institutional challenges. GEF moved out of the World Bank to become a separate and permanent institution with enhanced involvement from developing countries in decision making and implementation. A new governing structure was instituted, the first operating procedures ("the Instrument for the GEF")[4] were documented, and the funding cycle ("the GEF Replenishment") commenced. The World Bank took on the provision of the Trust Fund. The United Nations Development Program, the United Nations Environment Program, and other international organizations contributed to project development, management, and delivery.

Organizational Structure

International Agencies: GEF currently partners with 10 international agencies: the World Bank; the United Nations Development Program (UNDP); the United Nations Environment Program (UNEP); the United Nations Food and Agriculture Organization; the United Nations Industrial Development Organization; the African Development Bank; the Asian Development Bank; the European Bank for Reconstruction and Development; the Inter-American Development Bank; and the International Fund for Agricultural Development. Procedurally, the World Bank administers funding, UNDP oversees project development, and UNEP serves as the scientific and technical advisor. The remaining agencies contribute to the management and delivery of projects.

International Conventions: GEF is the primary fund administrator for four Rio (Earth Summit)[5] Conventions, including the Convention on Biological Diversity (CBD), the United Nations Framework Convention on Climate Change (UNFCCC), the Stockholm Convention on Persistent Organic Pollutants (POPs), and the United Nations Convention to Combat Desertification (UNCCD). GEF also establishes operational guidance for international waters and ozone activities, the latter consistent with the Montreal Protocol on Substances that Deplete the Ozone Layer and its amendments.

Internal Organization: GEF's main decision-making body is the GEF Council, which is an independent board of governors responsible for developing, adopting, and evaluating operational policies and programs. The Council is composed of 32 appointed members—16 from developing countries, 14 from developed countries, and 2 from among the countries of Central and Eastern Europe and the former Soviet Union. The balance between donor and recipient countries was negotiated and agreed to by Member countries after the pilot phase of the program. The Council meets approximately every six months and allows non-governmental organizations and private individuals to attend most sessions. Formal voting goes before the GEF Assembly, which is composed of representatives from all Member countries and meets every four years. During these times, the Assembly reviews general policy for operations, membership, funding, and amendments. The GEF Secretariat, based in Washington, DC, services and reports to the Council and the Assembly and formulates the work program, oversees implementation, and ensures that operational policies are followed.

Voting: The Assembly and the Council make decisions and adopt regulations through the process of *consensus*. GEF defines consensus as an agreement reached by all participants which includes the resolution or mitigation of all minority objections. If, in the case of the Council, all practicable efforts have been made and no consensus appears, Members may request a *formal vote*. The GEF formal vote is a double weighted majority; that is, an affirmative vote that includes both a 60% majority of the total number of Participants and a 60% majority of the total amount of contributions.[6] This format arose through a coordinated agreement between developed and developing countries in an effort to give facility to both donors and recipients in the decision-making process. A formal vote has never been taken at Council.

Funding

Replenishments: GEF is funded by donor countries, which pledge money every four years through a process known as GEF Replenishment. The process of Replenishment was designed to allow for program flexibility, strategic planning, and periodic performance evaluations. The original GEF pilot program of $1 billion has been replenished five times with $2.01 billion in 1994, $2.67 billion in 1998, $2.93 billion in 2002, $3.13 billion in 2006, and $4.34 billion in 2010.[7] Financial commitments by donor country to the GEF pilot program and the five GEF replenishments can be found in Figure A-1 of the Appendix.

U.S. Commitments: The United States supported the establishment of GEF in 1991. While the United States did not provide direct funding to the pilot phase of the program (1991-1993),[8] it has made commitments and contributions to all five GEF replenishments. U.S. commitments to the various four-year Replenishment cycles have been $430 million in 1994, $430 million in 1998, $430 million in 2002, $320 million in 2006, and $575 million in 2010, for a total of $2.185 billion. U.S. commitments correspond to 13.9% of total contributions for GEF during the history of the program, or, more specifically, 21.3% of total contributions for GEF-1, 16.1% for GEF-2, 14.7% for GEF-3, 10.2% for GEF-4, and 13.2% for GEF-5.[9]

U.S. Contributions: Payments made by the U.S. Treasury to the International Bank for Reconstruction and Development (the World Bank) as trustee for GEF have varied widely over the years due mainly to budget trends. Recent contributions include the following:

- For FY2010, P.L. 111-117 was enacted in December 2009 with a budget authority of $86.5 million to GEF, of which $80 million was for the final GEF-4 contribution and $6.5 million was for a portion of arrears.
- For FY2011, the U.S. Budget for Fiscal Year 2011, released in February 2010, had originally set forth a U.S. commitment of $680 million for the Fifth
- Replenishment, to be paid in four equal installments of $170 million from FY2011 through FY2014. The FY2011 Budget had included $170 million for the first installment of GEF-5 and $5 million for a portion of U.S. arrears, for a total request of $175 million.[10] However, in reaction to lower-than-expected pledge levels by other donor countries during the GEF-5 negotiations in May 2010, the U.S. delegation reduced its pledge to $575 million, to be paid in four equal installments of $143.75 million from FY2011 through FY2014. The U.S. pledge represents an 80% increase over the GEF-4 Replenishment. The increased pledge was precipitated in part by the U.S. Administration achieving certain policy reforms designed to improve GEF's overall effectiveness, particularly with regard to country-owned business plans for funding and resource allocation. P.L. 112-10 was enacted in April 2011 with budget authority of $89.82 million to GEF, some $53.93 million in arrears of the FY2011 pledge.

- For FY2012, P.L. 112-74 was enacted in December 2011 with a budget authority of $89.82 million to GEF, some $53.93 million in arrears of the FY2012 pledge. However, P.L. 112-74 included a provision under "Economic Support Fund" "that in consultation with the Secretary of the Treasury, the Secretary of State may transfer up to $200,000,000 of the funds made available under this heading to funds appropriated in this Act under the headings 'Multilateral Assistance, Funds Appropriated to the President, International Financial Institutions' for additional payments to such institutions, facilities, and funds enumerated under such headings." The Administration used this provision to transfer an additional $30 million to the GEF in FY2012.
- For FY2013, P.L. 113-6 was enacted in March 2013 with a budget authority of $129.4 million to GEF, some $14.35 million in arrears of the FY2013 pledge.
- For FY2014, the U.S. Budget for Fiscal Year 2014, released in April 2013, requested $143.75 million for the FY2014 contribution to GEF, equivalent to the FY2014 pledge.

Arrears: As of September 2012, direct payments to the trustee of GEF have totaled close to $1.8 billion over the past two decades. Further, in July 2011, the United States had cleared a portion of its GEF-3 arrears in the amount of $11.9 million through a retroactive early encashment of its GEF-4 contributions.[11]

With these payments, the United States is currently $258.9 million in arrears of its pledged commitments. As of the latest "Trustee Report" furnished to GEF in November 2012, the United States is joined in arrears status by three other countries: Egypt ($0.8 million), Nigeria ($1.0 million), and Spain ($3.1 million).[12]

A summary of U.S. commitments and contributions to GEF is shown in Table 1. The financial status of the GEF Trust Fund and the summary of arrears by country can be found in Figure A-2 of the Appendix.

Congressional Jurisdiction: All U.S. funding is subject to annual congressional approval. Authorizing legislation is managed by the House Financial Services Committee and Senate Foreign Relations Committee. The House and Senate Appropriations Subcommittees on State, Foreign Operations, and Related Programs have jurisdiction over appropriations.

Table 1. U.S. Commitments and Contributions to GEF by Fiscal Year

Fiscal Year	$ Committed to GEF ($ in millions, nominal)	$ Contributed to GEF ($ in millions, nominal)
1994	$0	$30.0
1995	$107.5	$90.0
1996	$107.5	$35.0
1997	$107.5	$35.0
1998	$107.5	$47.5
1999	$107.5	$167.5
2000	$107.5	$35.8
2001	$107.5	$107.8
2002	$107.5	$100.5
2003	$107.5	$146.8
2004	$107.5	$138.4
2005	$107.5	$106.7
2006	$107.5	$79.2
2007	$80.0	$79.2
2008	$80.0	$81.1
2009	$80.0	$80.0
2010	$80.0	$86.5
2011[a]	$143.8	$89.8
2012	$143.8	$119.8
2013[b]	$143.8	$129.4
2014	$143.8	TBD
Total	$2,185.2	$1,786.0

Source: CRS correspondence with U.S. Department of the Treasury.

[a] The enacted FY2011 figure includes the 0.2% rescission across all non-defense accounts, in accordance with Section 1119(a) of P.L. 112-10.

[b] The enacted FY2013 figure does not include sequestration reductions.

PROJECT AREAS

GEF funding is provided to recipient countries for projects and programs in six areas: biodiversity, climate change, international waters, ozone layer depletion, land degradation, and persistent organic pollutants. For examples of the types of projects funded by GEF, see the text box below.

Biodiversity: GEF is the financial mechanism of the 1992 United Nations Convention on Biological Diversity. The goal of GEF's program is the conservation and sustainable use of biodiversity, the maintenance of the ecosystem goods and services that biodiversity provides to society, and the fair and equitable sharing of the benefits arising out of the use of genetic resources. To achieve this goal, the program has several objectives including sustainability initiatives in protected areas, conservation measures in production sectors, capacity building to implement the Cartagena Protocol on Biosafety (CPB), and capacity building to support the implementation of the Bonn Guidelines on Access to Genetic Resources. Biodiversity projects constitute the largest percentage of GEF's portfolio, making up 36% of total grants.

Climate Change: As the financial mechanism of the 1992 United Nations Framework Convention on Climate Change, GEF allocates and disburses funding for projects in climate change mitigation (i.e., reducing or avoiding greenhouse gas emissions in the areas of renewable energy, energy efficiency, and sustainable transport), and climate change adaptation (i.e., increasing resilience to the adverse impacts of climate change of vulnerable countries, sectors, and communities). GEF projects in climate change help developing countries contribute to the overall objective of the UNFCCC to achieve a "stabilization of greenhouse gas concentrations in the atmosphere at a level that would prevent dangerous anthropogenic interference with the climate system." Moreover, GEF manages two special funds under the UNFCCC—the Least Developed Countries Fund, to assist in adaptation strategies for the most vulnerable countries; and the Special Climate Change Fund, to assist in mitigation and adaptation programs for countries that are heavily reliant of fossil-fuel technologies.

International Waters: GEF's international waters focal area does not serve as a financial mechanism for a specific convention. Through an association with regional agreements, it targets trans-boundary water systems, such as river basins with water flowing from one country to another, groundwater resources shared by several countries, and marine ecosystems bounded by more than one nation. GEF grants help countries collaborate with their neighbors to modify human activities that place stress on trans-boundary water systems and interfere with downstream uses of those resources. Some of the issues addressed include trans-boundary water pollution, over-extraction of groundwater resources, unsustainable exploitation of fisheries, control of invasive species, and balancing the competing uses of water resources.

Ozone Layer Depletion: GEF, in partnership with the 1985 Vienna Convention for the Protection of the Ozone Layer and the 1987 Montreal Protocol on Substances that Deplete the Ozone Layer, has aimed to safeguard the earth's protective ozone layer after the discovery that certain compounds were found to deplete it, posing substantial risks to human health and the environment. GEF has allocated funds to assist in phasing out ozone-depleting substances (ODS) and curbing the rising production and use of hydrochlorofluorocarbons (HCFCs). GEF's aim is to protect human health and the environment by assisting countries in phasing out consumption and production of ODS while enabling alternative technologies and practices, according to countries' commitments under the Montreal Protocol. The long-term goal of GEF interventions is to contribute to the return of the ozone layer to pre-1980 levels.

Examples of GEF Projects

1. Rural Electrification and Renewable Energy Development in Bangladesh (GEF ID 1209) GEF Grant: $8,540,000

Description: The project promoted solar energy in rural areas by (1) increasing awareness of Solar Heating Systems (SHS) among consumers and providers; (2) building technical and management capacity; (3) implementing and evaluating SHS programs; (4) providing technical and business development support to institutions; (5) introducing standards and programs for testing and certification; (6) financing grants to buy-down capital costs to increase affordability of SHS; (7) promoting electricity as a means for income generation and social wellness; and (8) identifying mechanisms to promote sustainability. Multiple approaches to SHS delivery were enacted, including a "fee-for-service" program through rural electricity cooperatives, purchase supported by micro-credit through NGOs and microfinance lenders, and hire-purchase/direct sale programs by private dealers and NGOs. Over 40,000 systems were installed supplying energy to rural dispersed communities.

2. Biodiversity Conservation in Cacao Agro-Forestry in Costa Rica (GEF ID 979) GEF Grant: $750,000

Description: The project improved management of cacao-based indigenous small-farms according to both ecological and organic productive principles so as to ensure conservation and sustainable use of plant and animal diversity and provide a sustainable source of family income.

The project promoted and maintained on-farm biodiversity while improving livelihoods of organic cacao producers (including indigenous, Latin-mestizos, and Afro-Caribbean groups) in the Talamanca-Caribbean corridor in Costa Rica.

3. Prevention and Management of Marine Pollution in the East Asian Seas in Indonesia (GEF ID 396) GEF Grant: $8,025,000

Description: The project developed policies and plans to control marine pollution from land and sea-based sources, upgraded national and regional infrastructures and technical skills, and established financing instruments to project sustainability. Project included selection of demonstration sites, establishment of regional monitoring and information networks, and involvement of regional association of marine legal experts to improve capacity to implement relevant conventions.

Source: GEF Project database at http://www.thegef.org/gef/gef_projects_funding (accessed November 30, 2011).

Notes: As of May 24, 2013, there were 3,404 projects listed in the GEF database, of which 2,783 were approved national projects and 621 were approved regional and global projects.

Land Degradation: In 2002, the GEF Assembly expanded GEF's mandate by adding land degradation to the portfolio and designating it the financial mechanism of the United Nations Convention to Combat Desertification. GEF focuses on sustainable agricultural practices (e.g., crop diversification, crop rotation, water harvesting, and small-scale irrigation schemes), sustainable rangeland management, and the preservation of viable indigenous forests and woodlands. GEF projects aim to integrate sustainable land management into national development priorities, and to strengthen human, technical, and institutional capacities

Persistent Organic Pollutants: GEF is the interim financial mechanism of the 2001 Stockholm Convention on Persistent Organic Pollutants, a global and legally binding agreement to reduce and eliminate pollutants including pesticides (e.g., DDT and mirex) and industrial chemicals (e.g., PCBs) as well as unintentionally produced POPs (e.g., dioxins and furans). GEF's involvement in tackling the threats posed by POPs dates back to 1995, with the introduction of the International Waters Operational Strategy and its contaminant-based component. In this framework, GEF began to develop a portfolio of strategically designed projects including regional assessments and pilot demonstrations that addressed a number of pressing POPs-related issues.

CURRENT ISSUES

Each year, billions of dollars in environmental aid flow from developed country governments— including the United States—to developing ones. GEF is one mechanism in the larger network of international programs designed to address environmental issues. While the efficiency and the effectiveness of these programs are of concern to donor country governments, a full analysis of the purposes, intents, results, and consequences behind these financial flows has yet to be conducted.[13] International relations, comparative politics, and developmental economics can often collide with environmental agendas. Critics contend that the existing system has had limited impact in addressing major environmental concerns—specifically climate change and tropical deforestation—and has been unsuccessful in delivering global transformational change. A desire to achieve more immediate impacts has led to a restructuring of the Multilateral Development Banks' role in environmental finance and the introduction of many new bilateral and multilateral funding initiatives.

The effectiveness of GEF depends on how the fund addresses its programmatic issues, reacts to recent developments in the financial landscape, and responds to emerging opportunities. The future of GEF remains in the hands of the donor countries that can choose to broaden the mandate and strengthen its institutional arrangements or to reduce and replace it by other bilateral or multilateral funding mechanisms. The following section investigates some of the current strengths and challenges facing GEF and summarizes some of the responses initiated by the program.

External Challenges for GEF

- *Rise of Climate Change Issues and Funding:* The past decade has seen a rise in the significance of global environmental issues—particularly climate change—on the political agendas of many countries. Proposed policies have not only attempted to address the environmental implications of greenhouse gas mitigation and climate change adaptation, but have become linked to energy and infrastructure issues through international economic, trade, and geo-political concerns. To address these issues, governments have begun to incorporate many global environmental objectives into their sustainable growth and development strategies. Funding for these activities has increased, and various institutional responses for this extensive portfolio are under consideration. Some critics contend that existing environmental funds (e.g., GEF) are unsatisfactory because they do not have experience managing investments of this scope.

- *The Changing Role of Multilateral Development Banks in Environmental Funding:* Multilateral Development Banks (MDBs) are key actors in the global system of environmental financing. As commercial lending institutions, some have argued that they dispense funds more efficiently than many institutional programs such as GEF; but as primary mechanisms for economic development, their past environmental lending practices have produced perceived conflicts of interest.[14] Objectives began to shift in 2005 when MDBs were encouraged by the G-8 leaders to play a more leading role in sustainable development and environmentally friendly technologies.[15] Since this time, MDBs have launched many new initiatives to address the environment, including efforts to (1) account for GHG emissions and improve energy efficiency; (2) support renewable energy; (3) manage forests sustainably; (4) promote carbon finance; and (5) adapt to climate change.[16] GEF programs now find themselves in competition with many of the new initiatives in MDB portfolios.
- *Increase in New Bilateral, Multilateral, and Private Funding Mechanisms:* Many donor governments perceive that the existing environmental finance system has not produced satisfactory results.[17] In searching for new and effective approaches to environmental funding, donors have sought options that can be organized quickly, administered directly, and be demonstrated to produce a more significant impact on the environment. Many have turned to highly specified multilateral programs, bilateral or even private sector measures to accomplish these aims, and no fewer than 15 environmental finance mechanisms have been announced since 2007. GEF is in competition with many of these new initiatives for a share of environmental funding.[18]

Internal Challenges for GEF

- *Low Level of Funding by Donor Countries:* Donor countries never intended GEF to cover all the financing needed to achieve developmental objectives. Rather, it was designed to be a catalyst for additional measures to address global environmental problems. As such, historical funding provided by donor countries was never at the level required to produce significant progress in reversing global threats. This experience has demonstrated that the initial assumptions underlying GEF—that relatively small amounts of incremental grant financing could leverage multilateral investment for transformational change—may be flawed.
- *Limits of Grant-based Instruments:* GEF was set up mostly to finance grants. Grants have proven to be inefficient in many development contexts given the greater leveraging and enhanced financial sustainability obtained from loan-based instruments. Such loans also provide reflows which can be lent again.
- *Difficulties in Defining "Incremental" and "Additional":* As stipulated in the "GEF Instrument," grants cover the "incremental" or "additional" cost of "transforming a project with national benefits into one with global environmental benefits." Incremental cost calculations have also been used as preference in project selection. Historically, GEF's implementing agencies have had difficulty producing a coherent methodology for calculating incremental cost, slowing the rate of project development. Furthermore, countries continue to argue over the requirements of additionality (i.e., whether or not the global environmental elements of a project would have taken place in the absence of GEF funding).
- *Difficulties with Adaptation:* GEF was established to finance global environmental benefits, which has made it difficult to justify GEF financing of climate change adaptation projects, which moreover provide local benefits.
- *Inefficient Procedures and Legal Status:* GEF's two-layer structure means that all funding must be approved twice (by GEF and the relevant GEF Agency), leading to inefficiencies. GEF's lack of legal status (the trust is held by the World Bank) prevents it from disbursing funding directly to countries with a one-step approval process.

- *Slowness of GEF Project Initiation and Implementation:* Since its inception, most note that GEF's project approval process has been long and complex. A 2006 internal report found a 66-month lapse between entry of a concept into the project pipeline and its initiation. Significant effort has been exerted to reduce the duration of the approval process, and the interval currently stands at 16 to 22 months. Bureaucratic structures, work program frequencies, Council deliberations, and consensus politics have all factored into delays.
- *Lack of Strategic Approach:* Historically, GEF has adhered to a project-byproject approach to allocating funds, wherein over 95% of pledges have been allocated to individual countries and less than 5% have been set aside for regional or global programs. The dynamic assumes that ongoing negotiations and incremental adjustments could foster transformational change in economies over time. While a project-by-project approach has allowed GEF to fulfill the mandates of many of its conventions, large-scale environmental issues such as climate change and biodiversity may demand more strategic and programmatic funding modalities.
- *Unsuccessful History of Leveraging the Private Sector:* While GEF has long recognized a need to mobilize investment resources in the private sector, successful collaboration may require a degree of experience and commitment that GEF cannot achieve under its existing structure. The length and uncertainty inherent in the GEF project cycle may make participation less attractive to the private sector, and the organization's emphasis on government entities at the expense of forming relationships with investors and manufacturers may serve as a further impediment.

GEF Reforms

GEF 4

During the 2006 Replenishment meetings, GEF worked to address many of its program deficiencies. The Council aimed to streamline costs and management fees, ensure project quality upon proposal, and reduce the length of the project pipeline.

A *Sustainability Compact* was enacted that would oversee several issues, including (1) the shift away from a project-oriented approach to a strategic and programmatic one; (2) a concentration on financing pre-market innovation in an attempt to leverage private capital; (3) a heightened dedication to transparency, accessibility, and equitability; and (4) a renewed focus on country-driven ownership through the implementation of a Resource Allocation Framework (RAF) wherein funding is determined by a country's potential to generate global environmental benefits and its capacity to successfully implement GEF projects. Further, in 2007, GEF initiated a pilot public-private partnership (PPP) initiative called the "Earth Fund" to enhance engagement with the private sector. Internal assessment of these reforms has shown promise.[19]

GEF 5

Meetings leading up to the Fifth Replenishment of GEF in 2010 saw the development of policy recommendations along two lines:

1. Enhancing Country Ownership: A key finding in GEF's recent performance evaluation was the relationship between country-driven strategic development and project success rate. Recommendations to strengthen country ownership include (1) reforming in-country corporate programs to include greater project portfolio identification and enhanced stakeholder coordination, (2) developing a more flexible and transparent resource allocation framework, and (3) broadening access to the GEF partnership to include national development agencies in developing countries.[20]
2. Improving the Effectiveness and Efficiency of GEF Partnerships: Recommendations to strengthen GEF partnerships include (1) enhancing accountability to the conventions and protocols; (2) streamlining the project cycle and refining the programmatic approach; (3) enhancing engagement with the private sector; (4) implementing the results-based management framework; (5) clarifying the roles and responsibilities of GEF entities, agencies, and conventions; and (6) enhancing engagement with civil society organizations.

GEF 6

Meetings for the Sixth Replenishment of GEF began on April 3, 2013, and continue on September 10, 2013. Policy recommendations are currently under development.

APPENDIX. GLOBAL ENVIRONMENT FACILITY TRUST FUND

Instrument for the Establishment of the Restructured
Global Environment Facility

GLOBAL ENVIRONMENT FACILITY TRUST FUND
PILOT PHASE

COMMITMENTS AS OF APRIL 30, 2007 **
(IN MILLIONS)

Contributing Participants	Pilot Phase	Total Contributions		
	SDR	Contribution[a]	Currency	
Australia	9.68	16.40	AUD	
Austria	26.02	29.07	EUR	e
Belgium	5.00 b	5.00	SDR	c
Brazil	4.00	4.00	SDR	c
Canada	6.33	10.00	CAD	
China	4.00	4.00	SDR	c
Côte d'Ivoire	2.00	2.00	SDR	c
Denmark	16.25	16.25	SDR	c
Egypt	4.00	4.00	SDR	c
Finland	20.44	17.66	EUR	e
France	110.08	123.00	EUR	e
Germany	110.02	122.85	EUR	e
India	4.00	4.00	SDR	c
Indonesia	4.00	4.00	SDR	c
Italy	65.14	54.23	EUR	e
Japan	27.36	5,373.00	JPY	
Mexico	4.00	5.48	USD	
Netherlands	37.74	37.74	SDR	c
Nigeria	4.00	4.00	SDR	c
Norway	19.56	165.00	NOK	
Pakistan	4.00	4.00	SDR	c
Portugal	4.50	4.50	SDR	c
Spain	10.00	10.00	SDR	c
Sweden	24.54	196.07	SEK	
Switzerland	30.06	30.06	SDR	c
Turkey	4.00	4.00	SDR	c
United Kingdom	54.73	40.30	GBP	
1. New Funding from Donors	615.45			
2. Contribution from IBRD Net Income	20.92			
Total (1 + 2)	636.37 d			

** Based on Instruments of Commitments or Qualified Instruments of Commitment received by the trustee.

a Calculated by converting the SDR amount to currency of contribution using an average daily exchange rate over the period from July 1, 1990 to Sep. 30, 1990.

b Contributing Participants had the option of taking a discount or credit for acceleration of encashment and; (i) including such credit as part of their basic share; (ii) counting such credit as a supplemental contribution; or (iii) taking such discount against the national currency contribution. Belgium opted to take a discount against their national currency contribution.

c These Contributing Participants denominated their contributions in SDRs.

d This is equivalent to USD 871.83 million using the agreed reference exchange rates for the GEF Pilot Phase

e The contributions of these Contributing Participants that are member states of the European Economic and Monetary Union (EMU) were originally pledged in each of their legacy currencies. The Trustee converted these legacy currencies into EUR, the common currency for the EMU, in January 2000 at fixed conversion rates.

Figure A-1. (Continued)

Instrument for the Establishment of the Restructured Global Environment Facility

UPDATE TO GLOBAL ENVIRONMENT FACILITY TRUST FUND
FIRST REPLENISHMENT OF RESOURCES *

COMMITMENTS AS OF APRIL 30, 2007 **
(IN MILLIONS)

Contributing Participants	Calculated Basic Contributions		Supplemental Contributions	Additional Supplemental Contributions	GEF-1 Total Contributions			
	(%)[a]	SDR	SDR	SDR	SDR	Contribution[b]	Currency	
Argentina	...	3.57			3.57	5.00	USD	
Australia	1.46%	20.84			20.84	42.76	AUD	
Austria	0.90%	12.85	1.05	0.37	14.28	16.82	EUR	c
Bangladesh	...	2.00			2.00	2.00	SDR	d
Belgium	1.55%	22.13	0.73		22.86	27.27	EUR	c
Brazil	...	4.00			4.00	4.00	SDR	d
Canada	4.00%	57.10	4.68		61.78	111.11	CAD	
China	...	4.00			4.00	4.00	SDR	d
Cote d'Ivoire	...	4.00			4.00	4.00	SDR	d
Czech Republic	...	4.00			4.00	4.00	SDR	d
Denmark	1.30%	18.56	1.52	5.00	25.08	25.08	SDR	d
Egypt	...	4.00			4.00	4.00	SDR	d
Finland	1.00%	14.28	1.17		15.45	20.86	EUR	c
France	7.02%	100.21	2.05		102.26	122.98	EUR	c
Germany	11.00%	157.03	12.86	1.41	171.30	171.30	SDR	d
Greece	0.05%	0.71	2.86		3.57	5.00	USD	
India	...	6.00			6.00	6.00	SDR	d
Ireland	0.11%	1.57	0.14		1.71	2.08	EUR	c
Italy	5.30%	75.66			75.66	82.53	EUR	c
Japan	18.70%	266.95	21.86	7.14	295.95	45,698.09	JPY	
Korea	0.23%	3.28	0.72		4.00	4.00	SDR	d
Luxembourg	0.05%	0.71	3.29		4.00	4.00	SDR	d
Mexico	...	4.00			4.00	4.00	SDR	d
Netherlands	3.30%	47.11	3.86		50.97	50.97	SDR	d
New Zealand	0.12%	1.71	0.14	2.15	4.00	10.35	NZD	
Norway	1.42%	20.27	2.02		22.29	220.00	NOK	
Pakistan	...	4.00			4.00	4.00	SDR	d
Portugal	0.12%	1.71	0.14	2.15	4.00	4.45	EUR	c
Slovak Republic	–	4.00			4.00	4.00	SDR	d
Spain	0.80%	11.42	2.55		13.97	13.10	EUR	c
Sweden	2.62%	37.40	3.06	1.14	41.60	450.04	SEK	
Switzerland	1.74%	24.84	2.03	5.10	31.97	31.97	SDR	d
Turkey	...	4.00			4.00	4.00	SDR	d
United Kingdom	6.15%	87.79	7.19	1.06	96.04	89.55	GBP	
United States	20.86%	297.78	9.14		306.92	430.00	USD	

Instrument for the Establishment of the Restructured
Global Environment Facility

Contributing Participants	Calculated Basic Contributions		Supplemental Contributions	Additional Supplemental Contributions	GEF-1 Total Contributions		
	(%)[a]	SDR	SDR	SDR	SDR	Contribution[b]	Currency
New Funding from Donors	89.80%	1,329.49	83.05	25.52	1,438.07		
Total					1,438.07 *e*		

** Based on Instruments of Commitment or Qualified Instruments of Commitment received by the Trustee.

a GEF basic share percentages are calculated based on new donor funding required for this replenishment in the amount of SDR 1,427.5 million.
As agreed by the Contributing Participants in December 1992, the basic shares for IDA 10 were the beginning shares of all non-recipient donors to the GEF-1.

b Calculated by converting the SDR amount to the currency of contribution using an average daily exchange rate over the period from Feb 1, 1993 to Oct 31, 1993 as agreed by the Contributing Participants at the GEF-1 replenishment meeting.

c The contributions of these Contributing Participants that are member states of the European Economic and Monetary Union (EMU) were originally pledged in each of their legacy currencies. The Trustee converted these legacy currencies into EUR, the common currency for the EMU, in January 2000 at fixed conversion rates.

d These Contributing Participants denominated their contributions in SDRs.

e This is equivalent to USD 2.01 billion using the agreed reference exchange rates for the GEF-1.

Figure A-1. (Continued)

Instrument for the Establishment of the Restructured Global Environment Facility

UPDATE TO GLOBAL ENVIRONMENT FACILITY TRUST FUND SECOND REPLENISHMENT OF RESOURCES[1]

COMMITMENTS AS OF APRIL 30, 2007 **
(IN MILLIONS)

Contributing Participants	Calculated Basic Contributions		Adjustment Toward Full Funding	Additional Supplemental Contributions	GEF-2 Total Contributions		
	(%)[a]	SDR	SDR	SDR	SDR	Contribution[b]	Currency
Australia	1.46%	21.95	1.52		23.47	43.27	AUD
Austria	0.90%	13.53	0.94 c	0.23 c	14.70 j	16.80	EUR h
Belgium	1.55%	23.30	1.62		24.92	30.94	EUR h
Canada	4.00%	60.13	4.17	10.30 g	74.60	141.66	CAD
China	..	4.00 d		2.00 f	6.00 f	6.00	SDR i
Cote d'Ivoire	..	4.00 d			4.00	4.00	SDR i
Czech Republic	..	4.00 d			4.00	4.00	SDR i
Denmark	1.30%	19.54	1.36		20.90	193.16	DKK
Finland	1.00%	15.03	1.04		16.07	19.63	EUR h
France	7.02%	105.54			105.54	131.50	EUR h
Germany	10.66%	160.32			160.32	198.99	EUR h
Greece	0.05%	0.75	0.05 e	3.20 e	4.00	4.46	EUR h
India	..	4.00 d		2.56 f	6.56 f	323.83	INR
Ireland	0.11%	1.65	0.12 e	2.23 e	4.00	4.69	EUR h
Italy	4.39%	65.97			65.97 j	73.85	EUR h
Japan	18.70%	281.13	19.54		300.67	48,754.33	JPY
Korea	0.23%	3.46	0.24 e	0.30 e	4.00	4,933.67	KRW
Luxembourg	0.05%	0.75	0.05 e	3.20 e	4.00	4.97	EUR h
Mexico	..	4.00 d			4.00	4.00	SDR i
Netherlands	3.30%	49.61	3.44		53.05	53.05	SDR i
New Zealand	0.12%	1.80	0.13 e	2.07 e	4.00	8.31	NZD
Nigeria	..	4.00 d			4.00	4.00	SDR i
Norway	1.42%	21.35	1.48		22.83	228.32	NOK
Pakistan	..	4.00 d			4.00	4.00	SDR i
Portugal	0.12%	1.80	0.13 e	2.07 e	4.00	4.90	EUR h
Slovenia	..	1.00			1.00	1.00	SDR i
Spain	0.80%	12.03			12.03	14.81	EUR h
Sweden	2.62%	39.39	2.73		42.12	448.07	SEK
Switzerland	1.74%	26.16	1.81	4.00	31.97	64.38	CHF
Turkey	..	4.00 d			4.00	4.00	SDR i
United Kingdom	6.15%	92.46	6.40	2.37	101.23	85.25	GBP
United States	20.84%	313.35			313.35	430.00	USD

Instrument for the Establishment of the Restructured
Global Environment Facility

Contributing Participants	Calculated Basic Contributions (%)[a]	Calculated Basic Contributions SDR	Adjustment Toward Full Funding SDR	Additional Supplemental Contributions SDR	GEF-2 Total Contributions SDR	GEF-2 Total Contributions Contribution[b] Currency
1. New Funding from Donors	88.53%	1,364.00	46.76	34.53	1,445.30	
2. Carryover of GEF resources					500.63 *k*	
Total (1 + 2)					1,945.93 *l*	

** Based on Instruments of Commitment or Qualified Instruments of Commitment received by the Trustee.

a GEF basic share percentages are calculated based on new donor funding required for this replenishment in the amount of SDR 1,503.35 million. The basic shares, which are originally derived from the GEF-1 and largely maintained in the GEF-2, do not add up to 100%.

b Calculated by converting the SDR amount to the currency of contribution using an average daily exchange rate over the period from May 1, 1997 to Oct 31, 1997 as agreed by the Contributing Participants at the GEF-2 replenishment meetings.

c The Adjustment toward Full Funding and the Additional Supplementary Contribution are the result of encashing EUR 16.8 million on a five-year, rather than 10-year schedule.

d Represents the agreed minimum contribution level to the GEF-2.

e These Contributing Participants agreed to adjust their contributions upward to the minimum contribution level of SDR 4 million.

f China and India agreed to contribute more than the minimum contribution level of SDR 4 million.

g This represents a transfer of CAD 19.6 million (USDeq. 13.4 million) from a trust fund established by Canada during the pilot phase of the GEF.

h The contributions of these Contributing Participants that are member states of the European Economic and Monetary Union (EMU) were originally pledged in each of their legacy currencies. The Trustee converted these legacy currencies into EUR, the common currency for the EMU, in January 2000 at fixed conversion rates.

i These Contributing Participants denominated their contributions in SDRs.

j The SDR value of the contribution is enhanced as the result of an agreed accelerated encashment schedule.

k Represents the amount carried over to the GEF-2 valued on the basis of June 30, 1998 exchange rates.

l This is equivalent to USD 2.67 billion using the agreed reference exchange rates for the GEF-2.

Figure A-1. (Continued)

Instrument for the Establishment of the Restructured
Global Environment Facility

UPDATE TO GLOBAL ENVIRONMENT FACILITY TRUST FUND
THIRD REPLENISHMENT OF RESOURCES[1]

COMMITMENTS AS OF APRIL 30, 2007 **
(IN MILLIONS)

Contributing Participants	Calculated Basic Contributions		Supplemental Contributions	GEF-3 Total Contributions		
	(%)[a]	SDR	SDR	SDR	Contribution[b]	Currency
Australia	1.46%	27.60		27.60	68.16	AUD
Austria	0.90%	17.01	0.69 c	17.70 g	24.38	EUR
Belgium	1.55%	29.30	3.67 c	32.97 g	41.98	EUR
Canada	4.28%	80.91		80.91	158.94	CAD
China	..	4.00 d	4.44 c	8.44 f g	7.50	SDR h
Cote d'Ivoire	..	4.00 d		4.00	4.00	SDR h
Czech Republic	..	4.00 d	0.50 c	4.50 g	4.00	SDR h
Denmark	1.30%	24.58	3.37	27.95	298.18 c	DKK
Finland	1.00%	18.91	2.03	20.94	30.00 c	EUR
France	6.81%	128.84 c		128.84	164.00	EUR
Germany	11.00%	207.96	23.66	231.62	293.67	USD
Greece	0.05%	0.95	3.55 c e	4.50 g	5.73	EUR
India	..	4.00 d	3.99 c	7.99 f g	426.39	INR
Ireland	0.11%	2.08	2.42 c e	4.50 g	5.73	EUR
Italy	4.39%	82.99		82.99	118.90	EUR
Japan	17.63%	333.41 c		333.41	48,754.33	JPY
Korea	0.23%	4.35		4.35	5.51 c	USD
Luxembourg	0.05%	0.95	3.05 e	4.00	5.73	EUR
Mexico	..	4.00 d		4.00	5.07 c	USD
Netherlands	3.30%	62.39		62.39	62.39 j	SDR h
New Zealand	0.12%	2.27	1.73 e	4.00	12.14	NZD
Nigeria	..	4.00 d		4.00	4.00	SDR h
Norway	1.06%	19.96		19.96	228.32	NOK
Pakistan	..	4.00 d		4.00	4.00	SDR h
Portugal	0.12%	2.27	1.73 e	4.00	5.73	EUR
Slovenia	..	1.00	0.13 c	1.13 g	1.00	SDR h
Spain	0.80%	15.12		15.12	21.67	EUR
Sweden	2.62%	49.53	7.45	56.98	764.67	SEK
Switzerland	2.43%	45.94		45.94	99.07	CHF
Turkey	..	4.00 d		4.00	4.00	SDR h
United Kingdom	6.92%	130.82 c	19.09 c	149.91 g	117.83	GBP
United States	17.94%	339.15 i		339.15	430.00 i	USD

Instrument for the Establishment of the Restructured
Global Environment Facility

Contributing Participants	Calculated Basic Contributions		Supplemental Contributions	GEF-3 Total Contributions		
	(%)[a]	SDR	SDR	SDR	Contribution[b]	Currency
1. New Funding from Donors	86.07%	1,660.29	81.50	1,741.79		
2. Supplemental Contributions including Credits			12.50 [c j]	12.50		
3. Investment Income				104.47 [k]		
4. Carryover of GEF Resources				450.00 [l]		
Total (1 + 2 + 3 + 4)				**2,308.76** [m]		

** Based on Instruments of Commitment or Qualified Instruments of Commitment received by the Trustee.

a GEF basic share percentages are calculated based on new donor funding required for this replenishment in the amount of SDR 1,890.5 million. The basic shares, which are originally derived from the GEF-1 and were largely maintained in the GEF-2, do not add up to 100%.

b Calculated by converting the SDR amount to currency of contribution using an average daily exchange rate over the period from May 15, 2001 to Nov. 15, 2001, as agreed by the Contributing Participants at the May 7, 2001, GEF-3 replenishment meeting.

c Contributing Participants had the option of taking a discount or credit for acceleration of encashment and; (i) including such credit as part of their basic share; (ii) counting such credit as a supplemental contribution; or (iii) taking such discount against the national currency contribution. France and Japan opted to include the credit for accelerated encashment in their basic share. The United Kingdom chose to accelerate encashment of its basic and supplemental contributions. A credit for accelerated encashment was thus included in its basic share and its supplemental contribution. Austria, Belgium, China, Czech Republic, Greece, India, Ireland, and Slovenia have opted to include the credit for accelerated encashment as a supplemental contribution. Denmark, Finland, Korea, and Mexico opted to take a discount against their national currency contribution. Canada chose to accelerate encashment of its contribution but not to take either a discount or a credit.

d Represents the agreed minimum contribution level to the GEF-3.

e These Contributing Participants agreed to adjust their contributions upward to the agreed minimum contribution level of SDR 4 million.

f China and India agreed to contribute more than the agreed minimum contribution level of SDR 4 million.

g The SDR value of the contribution is enhanced as the result of the agreed accelerated encashment schedule as noted in footnote *c*.

h Those Contributing Participants denominated their contributions in SDRs.

i The United States pledged USD 500 million (representing a basic share of 20.86%) during the GEF-3 negotiations, of which USD 70 million was conditional upon achievement of the performance measures outlined in Schedule 1 to Attachment 1 of the GEF-3 Resolution. Such measures were not met.

j Represents (i) a credit from acceleration from Canada in the amount of SDR 10.13 million and (ii) a supplemental contribution from The Netherlands in the amount of SDR 2.37 million, bringing The Netherlands' total SDR contribution to SDR 64.76 million.

k The actual investment income earned on GEF resources during the GEF-3 commitment period (FY03 through FY06) was USD 132.46 million. This amount is converted to SDR using the agreed reference exchange rates for the GEF-3.

l Represents the amount carried over to the GEF-3 pursuant to paragraph 9 of Resolution No. 2002-0005, valued on the basis of June 30, 2002 exchange rates.

m This is equivalent to USD 2.93 billion using the agreed reference exchange rates for the GEF-3.

Figure A-1. (Continued)

Instrument for the Establishment of the Restructured Global Environment Facility

ATTACHMENT 1 TO RESOLUTION NO. 2006-0008

GLOBAL ENVIRONMENT FACILITY TRUST FUND
FOURTH REPLENISHMENT OF RESOURCES
TABLE OF CONTRIBUTIONS

CONTRIBUTIONS (IN MILLIONS)

Contributing Participants	GEF-4 Shares and Basic Contributions[a]		Supplemental Contributions	Adjustment Towards Full Funding	Total Contributions		
	(%)	SDR	SDR	SDR	SDR	Currency[b]	Currency
Australia	1.46%	24.43	6.61	-	31.04	59.80	AUD
Austria	0.90%	15.06	7.26 c	-	22.32	24.38	EUR
Belgium	1.55%	25.94	12.83 c	3.51	42.28	46.18	EUR
Canada	4.28%	71.62	17.57	-	89.20	158.94 c	CAD
China	-	4.00 d	3.10 c	-	7.10	9.51	USD
Czech Republic	-	4.00 d	0.68 c	-	4.68	142.89	CZK
Denmark	1.30%	21.75	11.68	1.32	34.75	310.00	DKK
Finland	1.00%	16.73	10.82 c	0.94	28.50	31.12	EUR
France	6.81%	71.28 f	57.42	-	128.70	188.71 c	USD
Germany	11.00%	115.05 f	86.08 e	-	201.14	295.00	USD
Greece	0.05%	0.84	4.41 c	-	5.25	5.73	EUR
India	-	4.00 d	2.72 c	-	6.72	9.00	USD
Ireland	0.11%	1.84	3.41 c	-	5.25	5.73	EUR
Italy	4.39%	73.46	-	-	73.46	87.91	EUR
Japan	17.63%	184.40 f	23.56	-	207.96	33,687.97	JPY
Korea	0.23%	3.85	0.62 c	-	4.47	6,142.97	KRW
Luxembourg	0.05%	0.84	3.16	-	4.00	4.79	EUR
Mexico	-	4.00 d	-	-	4.00	63.38	MXN
Netherlands	3.30%	55.22	19.47	-	74.70	89.38	EUR
New Zealand	0.12%	2.01	1.99	-	4.00	8.40	NZD
Nigeria	-	4.00 d	-	-	4.00	4.00	SDR g
Norway	1.44%	24.11	-	-	24.11	228.32	NOK
Pakistan	-	4.00 d	-	-	4.00	350.01	PKR
Portugal	0.12%	2.01	2.78	-	4.79	5.73	EUR
Slovenia	0.03%	0.50	3.88 c	-	4.38	1,146.20	SIT
South Africa	-	4.00 d	-	-	4.00	38.27	ZAR
Spain	1.00%	16.73	1.37	-	18.11	21.67	EUR

REPLENISHMENT

Instrument for the Establishment of the Restructured
Global Environment Facility

Contributing Participants	GEF-4 Shares and Basic Contributions[a] (%)	GEF-4 Shares and Basic Contributions[a] SDR	Supplemental Contributions SDR	Adjustment Towards Full Funding SDR	Total Contributions SDR	Total Contributions Currency[b]	Total Contributions Currency
Sweden	2.62%	43.84	24.70	7.66	76.20	850.00	SEK
Switzerland	2.26%	37.82	-	9.67	47.49	88.00	CHF
Turkey	-	4.00 *d*	-	-	4.00	4.00	SDR *g*
United Kingdom	6.92%	115.80	56.08	-	171.88	140.00	GBP
United States	20.86%	218.18	-	-	218.18	320.00	USD
New Funding from Donors	89.43%	1,175.34	362.22	23.10	1,560.66		
Projected Investment Income					250.91 *h*		
Projected Carryover of GEF Resources					325.67 *i*		
Total Projected Resources to Cover GEF-4 Work Program					2,137.23 *j*		

a The GEF-4 basic shares reflect those of the GEF-3 except for Switzerland, Spain, Norway and Slovenia.

b As agreed by the Contributing Participants at the June 9-10, 2005 GEF-4 replenishment meeting, the reference exchange rate to convert the SDR amount to the national currency will be the average daily exchange rate over the period from May 1, 2005 to October 31, 2005.

c Contributing Participants have the option of taking a discount or credit for acceleration of encashment and; (i) including such credit as part of their basic share; (ii) counting such credit as a supplemental contribution; (iii) including such credit as an adjustment to full funding or (iv) taking such discount against the national currency contribution. Austria, Belgium, China, Czech Republic, Finland, Greece, India, Ireland, Korea and Slovenia have opted to take the credit for accelerated encashment as a supplemental contribution. Canada and France have chosen to take a discount against their contribution.

d For those Contributing Participants that do not have a basic share, this represents the agreed minimum contribution of SDR 4 million.

e Germany will provide this supplemental contribution of SDR 86.08 million under the terms of the GEF-4 replenishment resolution. This contribution will be made in order to strengthen the GEF's ability to meet funding objectives and policy commitments of the GEF-4 agreement. Progress towards meeting these commitments will be assessed in the GEF-4 midterm reviews and taken into account by Germany.

f These contributions are calculated to reflect a replenishment share based on the contributions of several major donors.

g As agreed by Contributing Participants in the June 9-10, 2005 GEF-4 replenishment meeting, Contributing Participants experiencing an average annual inflation rate in their economies exceeding 10% over the years 2002-2004 will denominate their GEF-4 contributions in SDR.

h Investment income is projected using a USD 2 billion average cash balance and investment return of 4.6% per annum.

i This amount comprises arrears, deferred contributions, and paid-in but unallocated resources.

j This amount is equivalent to USD 3.13 billion using the agreed GEF-4 reference exchange rates.

Source: Instrument for the Establishment of the Restructured Global Environment Facility - March 2008, at http://www.thegef.org/gef/node/2552

Figure A-1. (Continued)

GEF-5 Contribution Table

Date: 28-Jun-10

Contributing Participants	Ccy	GEF-5 Basic Share, %	GEF-5 Actual Share, %	Total Contributions (in millions)		
				SDR	USDeq	NC
1	2	3	4	15	16	17
Australia	AUD	1.46%	2.29%	52.88	81.03	105.00
Austria	EUR	1.21%	1.74%	40.15	61.53	42.60
Belgium	EUR	1.55%	3.33%	77.05	118.07	78.00
Brazil	USD	0.00%	0.35%	8.00	12.26	12.26
Canada	CAD	4.28%	5.85%	135.17	207.13	238.40
China	USD	0.00%	0.42%	9.79	15.00	15.00
Czech Republic	CZK	0.00%	0.20%	4.60	7.05	116.91
Denmark	DKK	1.30%	2.09%	48.40	74.16	400.00
Finland	EUR	1.00%	2.43%	56.20	86.11	57.80
France	EUR	6.76%	8.40%	194.16	297.52	215.50
Germany	EUR	10.89%	13.53%	312.64	479.08	347.00
Greece	EUR	0.05%	0.19%	4.35	6.67	4.44
India	USD	0.00%	0.28%	6.39	9.80	9.00
Ireland	EUR	0.11%	0.24%	5.62	8.61	5.73
Italy	EUR	2.80%	3.59%	82.80	127.02	92.00
Japan	JPY	11.48%	14.26%	329.55	505.00	48,377.08
Korea	USD	0.17%	0.23%	5.33	8.16	7.50
Luxembourg	EUR	0.05%	0.17%	4.00	6.13	4.44
Mexico	MXN	0.00%	0.28%	6.53	10.01	124.30
Netherlands	EUR	2.60%	3.23%	74.69	114.45	82.90
New Zealand	NZD	0.12%	0.17%	4.00	6.13	9.92
Nigeria	NGN	0.00%	0.17%	4.00	6.13	921.93
Norway	NOK	1.34%	1.66%	38.47	58.94	376.00
Pakistan	PKR	0.00%	0.17%	4.00	6.13	499.64
Portugal	EUR	0.12%	0.17%	4.00	6.13	4.44
Russian Federation	USD	0.00%	0.31%	7.10	10.89	10.00
Slovenia	EUR	0.03%	0.20%	4.71	7.21	4.80
South Africa	ZAR	0.00%	0.19%	4.35	6.67	51.56
Spain	EUR	0.97%	1.20%	27.76	42.54	30.81
Sweden	SEK	2.29%	3.70%	85.43	130.91	1,015.00
Switzerland	CHF	2.16%	3.26%	75.41	115.56	124.93
Turkey	TRY	0.00%	0.17%	4.00	6.13	9.57
United Kingdom	GBP	6.93%	9.28%	214.43	328.60	210.00
United States	USD	13.07%	16.23%	375.23	575.00	575.00
Total New Donor Funding		72.76%	100.00%	2,311.29	3,541.77	

		SDR	USDeq
1	**Total New Donor Funding**	**2,311.29**	**3,541.77**
2	Projected Investment Income	73.13	112.00
3	Projected Carryover of GEF Resources	448.27	686.55
4	Total Agreed Replenishment (1+2+3)	2,832.69	4,340.32

Source: CRS correspondence with GEF.

Figure A-1. Commitments to GEF Pilot Phase and Replenishments.

Countributing Participant	Repl.	Currency	Arrears Amount	USD eq.
Egypt	GEF-1	SDR	0.53	0.82
United States	GEF-2	USD	134.97	134.97
Nigeria	GEF-3	SDR	0.67	1.03
United States	GEF-3	USD	16.15	16.15
Spain	GEF-5	EUR	2.41	3.11
United States	GEF-5	USD	107.86	107.86
Total				**263.93**

Source: GEF, "Global Environment Facility Trust Fund Financial Report," GEF/C.43/Inf.08, September 30, 2012.

Figure A-2. Financial Status of GEF Trust Fund: Summary of Arrears.

End Notes

[1] See Figure A-1 of the Appendix for a list of donor countries during each GEF funding, or "Replenishment," period.

[2] Information on GEF activities, organization, policies, and projects is available on its website, at http://www.thegef.org/gef/.

[3] A full overview and analysis of the history of GEF and environmental financing can be found in a number of source materials including book length studies by Inge Kaul and Pedro Conceição, The New Public Finance: Responding to Global Challenges, New York: Oxford University Press, 2006; Robert L. Hicks, Bradley C. Parks, J. Timmons Roberts, and Michael J. Tierney, Greening Aid?: Understanding the Environmental Impact of Development Assistance, New York: Oxford University Press, 2008; and several articles including Gareth Porter, Neil Bird, Nanki Kaur, and Leo Peskett, "New Finance for Climate Change and the Environment," WWF and the Heinrich Böll Foundation, 2008; Smita Nakhooda, Jon Sohn, and Kevin Baumert, "Mainstreaming Climate Change Considerations at the Multilateral Development Banks," World Resources Institute, 2005; and Smita Nakhooda, "Correcting the World's Greatest Market Failure: Climate Change and the Multilateral Development Banks," World Resources Institute, 2008.

[4] The "Instrument for the Establishment of the Restructured Global Environment Facility" is the officially adopted operating procedures of GEF. See documents at http://www.thegef.org/gef/node/2552.

[5] The United Nations Conference on Environment and Development (UNCED), also known as the Rio Summit, Rio Conference, Earth Summit, and Eco '92 was a United Nations conference held in Rio de Janeiro from June 3-14, 1992, in which 172 governments participated, with 108 sending their heads of state or government, and 2,400 representatives

of non-governmental organizations (NGOs), with 17,000 people at the parallel NGO "Global Forum."

6 For the purpose of voting power, total contributions consist of the actual cumulative contributions made to the GEF Trust Fund.

7 Replenishment figures calculated in Special Drawing Rights (SDRs) from respective currencies at time of pledge. Figures are nominal. GEF's fiscal year runs from July 1 to June 30.

8 During the GEF pilot phase, the United States had a separate co-financing arrangement administered by USAID.

9 The percentage of total contributions and the percentage of new donor funding are computed from different totals. U.S. percentage of new donor funding, including supplemental contributions, is 21.3%, 21.7%, 19.5%, 14.0%, and 16.2%, respectively. See Appendix, Figure A-1, for a full breakdown of U.S. contributions in relation to other sources.

10 Office of Management and Budget, The Budget of the United States Government, 2011.

11 "Early encashment" is a payment mechanism used by GEF that allows donors to account for interest made on their contributions.

12 See GEF, "Global Environment Facility Trust Fund Financial Report," GEF/C.43/Inf.08, September 30, 2012, at http://www.thegef.org/gef/council_document/gef-trust-fund-financial-report.

13 See Hicks, et al., op. cit., for an initial foray into a quantified analytic and a discussion on the metrics involved.

14 The development portfolios of most MDBs strongly emphasize a bias toward conventional fossil fuel power generation and infrastructure loans that often worked counter to environmental aims (e.g., the World Bank loaned more than $2.5 billion for conventional power projects in 2005 compared to $109 million for renewable energy or energy efficiency). See Gareth Porter, et al., op. cit., for further comments.

15 See the Gleneagles Plan of Action at the 2005 G-8 Meeting in Gleneagles, Scotland. It should be noted that the portfolios of many MDBs still retain significant provisions for conventional power and infrastructure projects as compared to most bilateral environmental aid, albeit with a greater ratio of renewable and efficiency resources than in the past. See Hicks, et al., op. cit.

16 For a full analysis of the rise of MDBs in environmental finance, see Smita Nakhooda, Correcting the World's Greatest Market Failure: Climate Change and the Multilateral Development Banks, World Resources Institute, 2008.

17 Statistics confirm these perceptions: as a point of comparison, the success rate for multilaterally funded environmental projects often pales in comparison to education, health, or infrastructure projects. Only 25% of World Bank-financed environmental projects during the years 2001-2003 received a "satisfactory" project outcome rating, compared to 100% for education, 86% for health, and 87% for infrastructure. See Hicks, et al., op. cit., p.6.

18 Recent initiatives include the Global Climate Change Alliance (GCCA) of the European Commission; the International Window of the Environmental Transformation Fund (ETF-IW) of the United Kingdom; the Spanish Millennium Development Goals (MDG) Fund; the Japanese Cool Earth Partnership; the German International Climate Initiative; the Norwegian Agency for Development Cooperation (NORAD) Rainforest Initiative; the Australian Global Initiative on Forests and Climate (GIFC); the German Life Web Initiative; the World Bank Forest Carbon Partnership Fund (FCPF); the GEF Tropical Forest Account (TFA); the World Bank Clean Technology Fund (CTF); the GEF-IFC Earth Fund; the World Bank Strategic Climate Fund (SCF) and Pilot Program for Climate Resilience (PPCR); the Kyoto Protocol Adaptation Fund; and the Copenhagen Accord

Green Fund. For an analysis and overview of these new programs, see Porter, et al., 2008, op. cit.

[19] See GEF's "Fourth Overall Performance Study" and "Policy Recommendations for the Fifth Replenishment of the GEF Trust Fund," February 12, 2010, p. 4, at http://www.thegef.org/gef/node/2483.

[20] These policy recommendations correspond to those highlighted in the U.S. Budget for Fiscal Year 2011 contributing to the U.S. pledge increase for the GEF-5 Replenishment, p. 862.

INDEX

A

access, 19, 21, 23, 37, 39, 46, 62
accessibility, 62
accountability, 18, 62
accounting, 9, 15, 44
accreditation, 34, 39
adaptation, ix, 12, 18, 19, 22, 28, 32, 35, 37, 38, 40, 46, 55, 58, 60
Adaptation Fund, ix, 8, 13, 28, 29, 36, 38, 46, 74
Africa, 10
agencies, xi, 3, 8, 13, 21, 34, 39, 43, 46, 48, 49, 50, 60, 62
agricultural sector, 7
agriculture, 15, 16, 43
Algeria, 7
appropriations, vii, x, 1, 3, 6, 25, 28, 30, 41, 42, 43, 53
Appropriations Act, viii, 2, 4, 5, 6
assessment, 62
assets, 8, 13
atmosphere, 55
authority(s), 4, 35, 36, 42, 44, 52, 53
avoidance, viii, 2, 7
awareness, 56
AWG, 32, 33, 44

B

Bangladesh, 14, 15, 56
banks, 20, 26
base, x, 48, 49
benefits, x, 18, 46, 48, 49, 55, 60, 62
bias, 74
Bilateral, 41, 43, 59
bilateral aid, 40
biodiversity, x, 47, 49, 54, 56, 61
biomass, 11
Bolivia, 14
bottom-up, 32
Brazil, 14, 38
breakdown, 74
Burkina Faso, 14

C

cacao, 56
Cambodia, 14, 16
capacity building, 12, 38, 55
carbon, viii, 2, 5, 7, 8, 10, 12, 18, 23, 25, 37, 41, 46, 59
Caribbean, 14, 16, 17, 56
catalyst, 60
central bank, 45
certification, 56
challenges, vii, viii, x, 1, 3, 4, 7, 12, 18, 25, 35, 45, 47, 49, 50, 58
Chile, 7, 11
China, 20, 36, 38, 46

Climate Investment Funds (CIFs), v, vii, viii, ix, 1, 2, 4, 5, 6, 8, 9, 11, 13, 15, 17, 18, 19, 20, 21, 22, 23, 24, 25, 28, 29, 36, 38, 41, 46
civil society, 8, 9, 11, 13, 19, 39, 45, 62
clean energy, 5, 23
clean technology, 7, 10
Clean Technology Fund (CTF), viii, 2, 4, 5, 6, 7, 8, 9, 10, 11, 12, 13, 74
climate, vii, viii, ix, x, xi, 2, 3, 5, 9, 11, 12, 13, 15, 16, 17, 18, 19, 21, 22, 23, 24, 25, 27, 28, 29, 30, 31, 33, 37, 38, 39, 40, 41, 42, 43, 44, 45, 46, 47, 48, 49, 54, 55, 58, 59, 60, 61
climate change, vii, ix, x, xi, 3, 5, 11, 17, 19, 21, 22, 23, 24, 27, 28, 29, 30, 31, 33, 37, 38, 39, 41, 42, 43, 47, 48, 49, 54, 55, 58, 59, 60, 61
climate-resilient development, viii, 2, 12, 18
CO_2, 26
coal, 10, 23, 24, 26, 49
collaboration, 8, 13, 61
Colombia, 7, 11
combustion, 26
commercial, vii, 1, 3, 23, 26, 59
communication, 49
communities, 12, 23, 55, 56
comparative advantage, 38
compatibility, 43
competition, 39, 59
competitive advantage, 42
complementarity, 37, 41
compliance, 44
composition, 33, 45
compounds, 55
conference, ix, 28, 30, 35, 46, 73
conflict, ix, 2
conflict of interest, ix, 2
Congo, 14
Congress, vii, viii, x, 1, 2, 3, 6, 24, 25, 27, 28, 30, 31, 42, 43, 46, 47
consensus, 8, 19, 23, 34, 51, 61
consent, 30, 31, 44
conservation, 10, 54, 56
Consolidated Appropriations Act, viii, 2, 4, 5, 6
consumers, 56
consumption, 56
contaminant, 57
controversial, 36
convention, 25, 31, 55
Convention on Biological Diversity (CBD), xi, 48, 50, 54
cooperation, x, 42, 47, 48
coordination, ix, 2, 8, 13, 20, 21, 62
cost, x, 9, 20, 28, 42, 43, 60
Costa Rica, 56
covering, 7
CPB, 55
crop, 43, 57
cycles, 52

D

database, 24, 56
DDT, 57
decision-making process, 51
deficiencies, 61
deforestation, xi, 12, 17, 43, 48, 58
degradation, x, 47, 49, 54, 57
demonstrations, 58
Denmark, 14, 22, 29, 32
Department of Defense, viii, 2, 6
Department of the Treasury, 3, 4, 45
developed countries, vii, viii, 2, 3, 19, 22, 51
developing countries, vii, viii, ix, x, 1, 2, 3, 5, 6, 7, 8, 12, 18, 19, 20, 22, 23, 28, 29, 30, 32, 36, 37, 38, 39, 40, 41, 43, 45, 47, 48, 49, 51, 55, 62
developing economies, 45
development assistance, vii, x, 1, 3, 9, 24, 28, 30, 49
development banks, viii, ix, 2, 3, 5, 7, 12, 24, 39
diffusion, 9
direct investment, 10
direct payment, 53
disaster, 15, 16, 17

disbursement, 46
discontinuity, 35
distribution, 10, 30
divergence, 32
diversification, 18, 57
diversity, 56
Doha, 34
donor countries, viii, ix, x, xi, 2, 6, 8, 13, 19, 20, 28, 29, 30, 41, 46, 47, 48, 49, 51, 52, 58, 60, 73
donors, 9, 15, 20, 22, 37, 38, 51, 59, 74
drawing, 23

E

Earth Summit, xi, 48, 50, 73
East Asia, 56
Eastern Europe, 44, 51
economic activity, 5
economic affairs, vii, 1, 3
economic development, ix, 2, 5, 19, 29, 59
economic growth, vii, 3, 19, 29, 49
economics, 17, 58
economies in transition, 44
ecosystem(s), vii, 3, 54, 55
education, 74
Egypt, 7, 10, 53
EIT, 44
electricity, 19, 56
emission, 9
energy, 7, 10, 11, 18, 23, 43, 46, 49, 55, 56, 58, 59, 74
energy efficiency, 7, 10, 11, 23, 43, 55, 59, 74
energy security, 7, 18
energy supply, 43
entrepreneurs, 19
environment, vii, viii, x, xi, 1, 2, 3, 5, 15, 16, 17, 26, 46, 47, 48, 56, 59
environmental degradation, vii, 3
environmental issues, 18, 58, 61
environmental protection, vii, ix, 2, 24
equity, 9, 15
Europe, 49
European Commission, 74
exchange rate, 9, 15, 24
executive branch, 42
expertise, ix, 2, 20, 21
exploitation, 55
exports, 19
externalities, 49
extraction, 55

F

family income, 56
farms, 56
FDI, 3
fear, 22
financial, vii, viii, ix, x, xi, 1, 2, 3, 5, 7, 8, 10, 12, 17, 18, 19, 20, 22, 23, 24, 25, 27, 28, 29, 31, 32, 33, 36, 37, 38, 39, 40, 41, 42, 43, 44, 45, 46, 47, 48, 53, 54, 55, 57, 58, 60, 74
financial institutions, viii, 1, 8, 25, 39, 43
financial intermediaries, 10
financial markets, 19
financial resources, vii, 3, 7, 22, 39
financial support, 3, 29
financial system, 19
Forest Investment Program (FIP), viii, 2, 5, 12, 14
fisheries, 55
flexibility, 51
food, 5
food security, 5
force, 31
foreign aid, x, 24, 28
foreign assistance, 29
foreign direct investment, 3
forest management, 43
formation, 33, 38
formula, 39, 46
France, 5, 9, 20
freedom, 39
funding, vii, x, xi, 1, 4, 5, 6, 7, 8, 9, 13, 14, 18, 22, 23, 24, 25, 34, 35, 37, 38, 39, 40, 46, 47, 48, 49, 50, 51, 52, 53, 54, 55, 56, 58, 59, 60, 61, 62, 73, 74

funds, viii, ix, x, 2, 4, 5, 6, 8, 9, 10, 13, 15, 18, 20, 21, 22, 25, 27, 28, 29, 38, 39, 40, 41, 42, 44, 46, 53, 55, 56, 59, 61

G

Green Climate Fund (GCF), v, vii, ix, x, 27, 28, 29, 32, 33, 34, 35, 36, 38, 39, 40, 41, 42, 43, 45
Global Environment Facility (GEF), v, vii, ix, x, xi, 4, 5, 8, 13, 21, 25, 28, 29, 36, 38, 41, 46, 47, 48, 49, 50, 51, 52, 53, 54, 55, 56, 57, 58, 59, 60, 61, 62, 63, 71, 72, 73, 74, 75
geo-political, 59
Germany, 5, 9, 14, 20, 22
GHG, 31, 32, 43, 45, 59
global climate change, vii, 1, 6, 43
global economy, 20, 45
global environmental problems, vii, 3, 60
global prosperity, 20
global scale, 21
goods and services, 10, 54
governance, viii, 2, 6, 7, 12, 20, 24, 35
governments, vii, x, 3, 8, 11, 17, 21, 22, 25, 31, 41, 47, 48, 58, 59, 73
grants, ix, x, 8, 9, 12, 13, 15, 19, 28, 29, 39, 47, 48, 55, 56, 60
greenhouse, viii, 2, 7, 9, 12, 55, 58
greenhouse gas, viii, 2, 7, 9, 12, 55, 58
greenhouse gas emissions, viii, 2, 7, 12, 55
Grenada, 14, 16
grid technology, 10
groundwater, 55
growth, 7, 18, 20, 22, 29
guidance, x, xi, 7, 23, 26, 28, 30, 33, 35, 36, 42, 45, 48, 50
guidelines, 36
Guinea, 14, 17

H

Haiti, 14, 17
harmonization, 21, 25
harvesting, 57
health, 56, 74
history, 23, 24, 52, 73
Honduras, 14
host, 30, 34
House, x, 6, 20, 28, 42, 46, 53
House of Representatives, x, 28, 42
human, vii, 3, 5, 31, 42, 43, 55, 57
human health, 5, 55
humanitarian aid, 40
hydroelectric power, 10

I

ID, 56
idealism, 49
identification, 62
images, 25
improvements, 43
income, ix, 11, 12, 27, 29, 31, 36, 38, 49, 56
India, 7, 11, 38
indigenous peoples, 8, 13
individuals, 51
Indonesia, 7, 11, 14, 56
industrial chemicals, 57
industrialized countries, ix, 3, 6, 27, 29, 43, 44
industry(s), 7, 9, 42, 43
infrastructure, 10, 11, 15, 16, 17, 19, 21, 23, 43, 46, 58, 74
initiation, 61
institutions, vii, x, 1, 3, 19, 28, 47, 48, 53, 56, 59
integration, 19
Inter-American Development Bank, 24, 50
interest rates, 25
interference, 55
International Bank for Reconstruction and Development, xi, 25, 48, 49, 52
international financial institutions, vii, x, 1, 3, 42, 47, 53
international meetings, 42
International Monetary Fund, 49
international standards, 44

investment(s), vii, viii, xi, 1, 2, 3, 5, 7, 8, 9, 10, 12, 13, 14, 18, 19, 21, 23, 46, 48, 59, 60, 61
investors, 61
irrigation, 43, 57
issues, vii, viii, ix, xi, 2, 3, 18, 21, 28, 30, 32, 38, 43, 44, 45, 48, 55, 58, 61

J

Jamaica, 14, 16
Japan, 5, 9, 14, 20
Jordan, 7
jurisdiction, 6, 53

K

Kazakhstan, 7, 11
Kenya, 14
Korea, 14, 34
Kyoto Protocol, 31, 32, 44, 46, 74

L

landscape, viii, x, xi, 2, 4, 15, 16, 17, 18, 41, 47, 48, 58
laws, 43
lead, vii, 3, 42
leadership, vii, 1, 3, 42
Least Developed Countries, 34, 55
legislation, 6, 43, 53
lending, 9, 19, 23, 40, 59
Liberia, 14
light, viii, x, 1, 4, 36, 41, 47
loan guarantees, 3, 8, 13
loans, 8, 9, 12, 13, 15, 19, 22, 25, 40, 60, 74
low-carbon technologies, viii, 2, 7

M

majority, 9, 36, 50, 51
management, 10, 12, 15, 16, 17, 25, 43, 46, 50, 56, 57, 61, 62
materials, 24, 73
membership, 34, 51
methodology, 60
Mexico, 7, 10, 14
Middle East, 7, 10
models, 45
modernization, 10
Montreal Protocol, xi, 3, 48, 50, 55
Morocco, 7, 10
Mozambique, 14, 16
Multilateral, xi, 4, 19, 24, 25, 41, 48, 50, 53, 58, 59, 73, 74
Multilateral Development Banks' (MDBs'), xi, 48, 58

N

national security, 42
natural gas, 10
natural resources, vii, 3
negotiating, 32, 33, 35, 36, 37, 38, 39, 40, 41, 44, 45
negotiation, 31, 44
Nepal, 14, 16
Netherlands, 14
NGOs, 8, 13, 21, 23, 56, 74
Nigeria, 7, 11, 53
North Africa, 7
North America, 45
Norway, 14, 22

O

Official Development Assistance (ODS), 24, 56
Organization for Economic Cooperation and Development (OECD), 24, 44
Office of Management and Budget, 4, 74
operations, 8, 12, 22, 34, 44, 51
opportunities, viii, xi, 2, 18, 20, 43, 48, 58
oversight, x, 28, 30, 42
ownership, 39, 62
ozone, x, 47, 49, 50, 54, 55
ozone layer, x, 47, 49, 54, 55

P

Pacific, 14, 16, 17
parallel, 21, 22, 74
participants, 51
PCBs, 57
Persistent Organic Pollutants, 57
personality, 34
Peru, 14
Philippines, 7, 10
Pilot Program for Climate Resilience (PPCR), viii, 2, 5, 12, 13, 15, 16, 17, 74
pipeline, 61
plants, 10, 26
Poland, ix, 28, 30, 35
policy, ix, 2, 12, 20, 32, 43, 49, 51, 52, 62, 75
policy reform, 12, 52
politics, 17, 58, 61
pollutants, x, 29, 47, 49, 54, 57
pollution, vii, 3, 29, 55, 56
POPs, 49, 57
portfolio, 23, 36, 55, 57, 58, 59, 62
poverty, vii, 3, 5, 7, 11, 19, 23, 29
poverty alleviation, 7, 19
poverty reduction, vii, 3, 5, 23
power generation, 10, 12, 23, 24, 26, 49, 74
power plants, 24
pragmatism, 49
prejudice, 22
preservation, 57
President, 31, 53
President Clinton, 31
principles, 7, 56
Private Funding, 59
private sector, viii, ix, x, 2, 5, 8, 10, 11, 13, 19, 22, 25, 28, 29, 30, 42, 46, 47, 49, 59, 61, 62
private sector investment, 42
privatization, 10
procurement, 20, 39
producers, 56
professionals, 21
programming, 6, 7, 13
project, x, 8, 18, 36, 40, 48, 49, 50, 56, 60, 61, 62, 74
property rights, 19
protected areas, 55
protection, 24
public finance, 37
public health, 43

R

rainfall, 11
rangeland, 57
ratification, 31, 43, 44
recommendations, 34, 62, 75
reform(s), 11, 62
regional integration, 18
regulations, 51
relevance, 25
renewable energy, 7, 11, 12, 23, 55, 59, 74
repackaging, 5
requirements, 9, 15, 43, 60
Residential, 11
resilience, 12, 29, 55
resolution, 4, 6, 44, 51
resource allocation, 19, 52, 62
resources, 5, 11, 16, 22, 23, 29, 32, 35, 37, 38, 39, 41, 55, 61, 74
response, 11, 43
restructuring, xi, 18, 48, 58
rhetoric, 23
risk(s), vii, 3, 8, 10, 12, 13, 15, 16, 17, 19, 21, 42, 55
risk management, 8, 13, 15, 16, 17
river basins, 55
rules, 19, 20, 36
rural areas, 56

S

Samoa, 14, 16
Scaling Up Renewable Energy Program (SREP), viii, 2, 5, 12, 14
Strategic Climate Fund (SCF), viii, 2, 4, 5, 6, 7, 11, 12, 13, 14, 24, 74

science, 43
scope, viii, 2, 6, 35, 38, 40, 59
SCP, 24
SDRs, 74
sea level, 11
sea-level, 43
sea-level rise, 43
Secretary of the Treasury, 53
seed, 10
Senate, x, 6, 28, 31, 42, 44, 53
Senate Foreign Relations Committee, 6, 53
services, 34, 35, 39, 43, 51
shape, 9
SIDS, 34
social welfare, 19
society, 19, 54
South Africa, ix, 7, 10, 28, 29, 33, 38, 45
South Korea, 34
sovereignty, 42
Soviet Union, 51
Spain, 9, 14, 53
species, 55
spending, 4, 44
spillover effects, 42
Spring, 45
stability, vii, 3
stabilization, 55
stakeholders, 21
state(s), 8, 21, 24, 25, 36, 38, 39, 73
state-owned enterprises, 8, 25
Stockholm Convention on Persistent Organic Pollutants (POPs), xi, 48, 50, 57
strategic planning, 51
stress, 40, 55
structure, viii, x, 1, 2, 4, 6, 7, 11, 12, 18, 19, 20, 21, 25, 31, 36, 37, 39, 47, 50, 60, 61
style, 50
success rate, 62, 74
support staff, 36
sustainability, 55, 56, 60
sustainable development, 5, 7, 24, 59
sustainable growth, vii, 3, 59
Sweden, 9, 14, 22
Switzerland, 14, 22, 34

T

Tajikistan, 14, 15
Tanzania, 14
target, 18, 44
technical assistance, ix, 12, 13, 27, 43
technology(s), viii, 2, 7, 9, 11, 21, 23, 25, 26, 32, 38, 49, 56
technology transfer, 32
temperature, 11
tensions, vii, 3
territory, 31
testing, 56
Thailand, 7, 10
threats, 57, 60
Tonga, 14, 17
top-down, 32
trade, vii, 1, 3, 39, 59
training, 19
transactions, 3
transformation, 18
transmission, 10
transparency, ix, 2, 18, 32, 62
transport, 10, 11, 55
transportation, 7, 43
Treasury, 4, 5, 20, 21, 44
treaties, 44
trust fund, viii, 2, 6, 8, 12, 13, 18, 20, 22, 50, 53, 63, 73, 74, 75
Turkey, 7, 10, 11

U

U.S. Department of the Treasury, 54
U.S. economy, 20, 44
U.S. involvement, vii, 1, 3
U.S. Treasury, 26, 52
Ukraine, 7, 11, 24
United Nations Development Program (UNDP), xi, 48, 49, 50
United Nations Framework Convention on Climate Change (UNFCCC), ix, x, xi, 3, 6, 7, 8, 13, 22, 25, 27, 28, 29, 31, 32, 33,

34, 35, 36, 38, 40, 41, 42, 43, 44, 45, 46, 48, 50, 55
United, vii, ix, x, xi, 1, 3, 4, 5, 9, 15, 17, 19, 20, 22, 24, 25, 27, 29, 31, 34, 36, 38, 40, 42, 43, 44, 45, 47, 48, 49, 50, 52, 53, 54, 55, 57, 58, 73, 74
United Kingdom, 5, 9, 15, 20, 22, 49, 74
United Nations, ix, xi, 3, 6, 22, 27, 29, 31, 34, 43, 44, 48, 49, 50, 54, 55, 57, 73
United Nations Convention to Combat Desertification (UNCCD), xi, 3, 48, 50, 57
United Nations Industrial Development Organization, 50
United States, vii, ix, x, xi, 1, 3, 4, 5, 9, 15, 17, 19, 20, 24, 25, 27, 29, 31, 36, 38, 40, 42, 44, 45, 47, 48, 49, 52, 53, 58, 74
urban, 5, 7, 16, 17, 23

V

varieties, 43
vehicles, 18
veto, 20
Vietnam, 7, 10
vote, 8, 44, 51
voting, 19, 20, 25, 51, 74

W

Washington, 5, 8, 13, 51
water, 7, 10, 11, 15, 16, 43, 55, 57
water heater, 7, 10
water resources, 15, 16, 43, 55
wellness, 56
wind power, 11
windows, 34, 39
World Bank, viii, 2, 4, 5, 8, 12, 13, 20, 21, 23, 24, 25, 26, 34, 36, 38, 43, 49, 50, 52, 60, 74

Y

Yemen, 14, 17